MAY 19.198_

D0319698

NO LONGER THE
PROPERTY OF
ELON UNIVERSITY LIBRARY

Thermodynamics:
Second Law Analysis

Thermodynamics:
Second Law Analysis

Richard A. Gaggioli, Editor

Marquette University

Based on a symposium sponsored
by the Division of Industrial
and Engineering Chemistry at
the 176th Meeting of the
American Chemical Society,
Miami, Florida,
September 11–14, 1978.

ACS SYMPOSIUM SERIES **122**

824825

AMERICAN CHEMICAL SOCIETY

WASHINGTON, D. C.   1980

Library of Congress CIP Data

Symposium on Theoretical and Applied Thermody-
namics, Miami, Fla., 1978.
Thermodynamics.
(ACS symposium series; 122 ISSN 0097–6156)

Includes bibliographies and index.

1. Thermodynamics—Congresses.
I. Gaggioli, Richard A. II. American Chemical Society. Division of Industrial and Engineering Chemistry. III. Title. IV. Title: Second law. V. Series: American Chemical Society. ACS symposium series; 122.

QC310.15.S95 1978   541.3'69   80–10486
ISBN 0–8412–0541–8   ACSMC8   122 1-301 1980

Copyright © 1980

American Chemical Society

All Rights Reserved. The appearance of the code at the bottom of the first page of each
article in this volume indicates the copyright owner's consent that reprographic copies of
the article may be made for personal or internal use or for the personal or internal use of
specific clients. This consent is given on the condition, however, that the copier pay the
stated per copy fee through the Copyright Clearance Center, Inc. for copying beyond that
permitted by Sections 107 or 108 of the U.S. Copyright Law. This consent does not extend
to copying or transmission by any means—graphic or electronic—for any other purpose,
such as for general distribution, for advertising or promotional purposes, for creating new
collective works, for resale, or for information storage and retrieval systems.

The citation of trade names and/or names of manufacturers in this publication is not to be
construed as an endorsement or as approval by ACS of the commercial products or services
referenced herein; nor should the mere reference herein to any drawing, specification,
chemical process, or other data be regarded as a license or as a conveyance of any right or
permission, to the holder, reader, or any other person or corporation, to manufacture, repro-
duce, use, or sell any patented invention or copyrighted work that may in any way be
related thereto.

PRINTED IN THE UNITED STATES OF AMERICA

ACS Symposium Series

M. Joan Comstock, *Series Editor*

Advisory Board

David L. Allara

Kenneth B. Bischoff

Donald G. Crosby

Donald D. Dollberg

Robert E. Feeney

Jack Halpern

Brian M. Harney

Robert A. Hofstader

W. Jeffrey Howe

James D. Idol, Jr.

James P. Lodge

Leon Petrakis

F. Sherwood Rowland

Alan C. Sartorelli

Raymond B. Seymour

Gunter Zweig

FOREWORD

The ACS SYMPOSIUM SERIES was founded in 1974 to provide a medium for publishing symposia quickly in book form. The format of the Series parallels that of the continuing ADVANCES IN CHEMISTRY SERIES except that in order to save time the papers are not typeset but are reproduced as they are submitted by the authors in camera-ready form. Papers are reviewed under the supervision of the Editors with the assistance of the Series Advisory Board and are selected to maintain the integrity of the symposia; however, verbatim reproductions of previously published papers are not accepted. Both reviews and reports of research are acceptable since symposia may embrace both types of presentation.

CONTENTS

PREFACE

The usefulness of direct application of the Second Law of Thermodynamics to the planning and engineering of energy conservation is being recognized finally; for example, see the 1978 National Energy Conservation Act, Section 683. The utility of the Second Law was amply shown at a recent workshop at George Washington University which occurred on August 14–16, 1979 and was sponsored by the Department of Energy.

Over the years prominent thermodynamicists have advocated Second Law analyses for properly evaluating energy-conversion processes on the basis of available energy (exergy). Available energy goes back to Maxwell and Gibbs. Unfortunately, it has not yet taken hold in engineering practice or in managerial decision making.

However, it is becoming increasingly clear in the energy-conversion literature that traditional gauges of energy efficiency are unsatisfactory. The reason is that the scientific concept of energy is assumed to be the commodity of value. (In science, the word "energy" is associated with the First Law of Thermodynamics, which says that no energy is consumed (used up) by processes.) Whereas, the true resource of value is the lay concept of energy—known as available energy (or exergy)—in the scientific and engineering literature. The Second Law of Thermodynamics says that exergy is the fuel that drives processes, and that it is consumed in doing so.

Various inconsistencies arise as a consequence of viewing (scientific) energy as the resource. Because energy cannot be consumed, whatever energy is supplied with fuel must end up somewhere—if not in the desired product, then in some waste. Consequently, effluent wastes are grossly overestimated in value while consumptions within processes—the major inefficiencies— are overlooked completely. For example, the usual home-heating furnace (or electric power plant boiler) appears to be very efficient (ca. 70%). For every 100 units of energy supplied with fuel, about 70 units go into the heated air and 30 units are lost with combustion gases discharged via the chimney. In actuality, such a furnace is only about 15% efficient; 30% of the fuel's exergy is consumed by the combustion process, which converts chemical exergy into thermal exergy. About 45% is consumed in the transfer of heat from the very hot products of combustion to the warm air; 10% is lost with the exhausted combustion gases. Thus, a total of 75%, not zero, is consumed, while about 10%, not 30%, is lost with the exhaust.

Frustrated by the inconsistencies associated with energy efficiencies, practitioners continue with new proposals of alternative definitions. Ironically, the vast majority persist in using energy as the measure of "potential to cause change." Consequently, the frustrations are destined to be perpetuated.

The key to resolving this dilemma simply is to recognize that exergy is the proper measure. With exergy analysis, which involves the same calculational procedures as energy analysis, the true inefficiencies and losses can be determined.

The concept of exergy is crucial not only to efficiency studies but also to cost accounting and economic analyses. Costs should reflect value; since the value is not in energy but in exergy, assignment of cost to energy leads to misappropriations, which are common and often gross. Using exergy content as a basis for cost accounting is important to management for pricing products and for their evaluation of profits. It is also useful to engineering for operating and design decisions, including design optimization.

Thus, exergy is the only rational basis for evaluating: fuels and resources; process, device, and system efficiencies; dissipations and their costs; and the value and cost of system outputs.

The chapters in this symposium volume illustrate the usefulness and develop the methodology of such Second Law analyses, now made much more comprehensible as a result of recent progress in Thermodynamics; survey the results of efficiency analyses of a variety of processes, devices, systems, and economic sectors; and teach the methods of engineering application of exergy to efficiency analysis and costing.

While baring many misconceptions resulting from energy analyses, the results of the efficiency analyses show great potential for alleviating the energy problem via conservation—even moreso over the intermediate and long term than over the short term—and pinpoint where the opportunities are. In turn, the cost analyses show how economic analysis decisions regarding energy systems can be facilitated greatly, while avoiding the misappropriations, which are often gross, that result from energy analyses.

The symposium volume will be valuable to energy and process engineers involved in design and in operating decisions, to managers in the private and government sectors who are involved with energy use and development, and to public service commissions.

Marquette University
1515 W. Wisconsin Ave.
Milwaukee, WI 53233

October 17, 1979

RICHARD A. GAGGIOLI

THERMODYNAMIC EVALUATION
OF PROCESS AND ENERGY
FACILITIES: EFFICIENCY ANALYSIS

Principles of Thermodynamics

RICHARD A. GAGGIOLI

Department of Mechanical Engineering, Marquette University,
1515 W. Wisconsin Ave., Milwaukee, WI 53233

This paper gives a simple, comprehensible presentation of
(a) the first and second laws of thermodynamics; (b) their associated basic concepts of energy and available-energy, respectively; and, (c) their practical implications on the performance of
processes and equipment. It will be seen that is is available
energy, not energy, which is the commodity of value and, hence,
the proper measure for assessing inefficiencies and wastes.

Thermodynamics - Its Basic Implications

The basic concepts of Thermodynamics are two commodities
called Energy and Available-Energy. The basic principles are the
First Law, dealing with energy, and the Second Law, dealing with
available-energy. (Different authors have presented the concept,
available-energy, with a variety of names: available-work, ener-
gy-utilisable, exergy, essergy, potential energy, availability,
. . .).
To illustrate the basic concepts and principles, picture a
conduit carrying some commodity such as electric charge, or high-
pressure water, or some chemical like hydrogen (H_2). The flow
rate of any such commodity is called a current and may be expressed as

I_q coulombs per second (amperes)

I_v gallons per minute

I_{H_2} moles per second

The conduit could be a heat conductor carrying a thermal current,
I_θ. Whatever the commodity might be, energy is transported con-
currently with it. The rate, I_E, at which energy flows is proportional to the commodity current. Thus, with charge current,

0–8412–0541–8/80/47–122–003$05.00/0
© 1980 American Chemical Society

I_q, the electric flow rate of energy past a cross-section of the conduit is

$$I_E = \phi I_q$$

where ϕ is the local value of the electric potential at that cross-section.

Likewise, the hydraulic energy flow rate associated with the volumetric current, I_v, is

$$I_E = p I_v$$

where p is the pressure. When a material flows and carries energy not only because of its pressure but also because of its composition, the flow of energy can be called a chemical flow

$$I_E = \mu_{H_2} I_{H_2}$$

where μ_{H_2} is the chemical potential.

Notice that, in each of the above examples, the proportionality factor between the commodity current and the associated energy current turns out to be the "potential" which drives the commodity through the conduit. (Stated more precisely, the potential gradient causes the flow.)

The driving force which causes a thermal current is a temperature difference, and the flow rate of energy with thermal current is given by

$$I_E = T I_\theta$$

Traditionally, in science and engineering, it is the flow rate of energy, I_E, that has been called the rate of <u>heat flow</u>. It would have been better to use the word "heat" (or "heat content") for the commodity flowing with current I_θ, but this commodity was not recognized until later, and has been named <u>entropy</u>. (Obert (<u>1</u>) introduced entropy as that commodity with which heat transfers of energy are associated, with temperature T as the proportionality coefficient -- in analogy with p as the proportionality coefficient between energy and volume transfers (or ϕ as that between energy and charge transfers). Much of the perplexity which thermodynamics has had is a result of insisting on providing a mathematical derivation of entropy from other concepts -- like "heat" ("heat energy") and temperature -- in contrast to simply providing <u>motivation</u> that it exists, as is done for its analog, charge.)

<u>Commodity Balances.</u> In analysis of energy converters, <u>balances</u> are applied for each of the relevant <u>commodities</u>; for examples, mass balances, energy balances, chemical compound balances, and so on. The amount of any given commodity in some container

can in general be changed either (1) by transporting the commodity
into or out of the container, or (2) by production or consumption
inside. Thus, on a rate basis

The rate of change in the amount of the commodity contained	= The sum of all the inlet rates	- The sum of the outlet rates

+ The rate of Production inside	-	The rate of consumption inside

For <u>steady</u> operation the rate of change in the amount of commodity
contained within the device or system is equal to zero.

Some commodities, like charge, that cannot be produced or
consumed, are said to be <u>conserved</u>.

<u>The First Law of Thermodynamics</u> states that

    (1)   Energy is conserved.
    (2)   The transport of any commodity has an associated
           energy transport.

<u>The Potential to Cause Change for Us: A Commodity</u>. When
does a commodity have the capacity to cause changes for us? The
answer is: whenever it is not in complete, stable equilibrium
with our environment. Then, it can be used to accomplish any kind
of change we want, to some degree. Thus, charge has this capacity
whenever it is at a potential different from "ground;" water has
this capacity whenever it is at a pressure different from "ground".
Several examples are illustrated in Fig. 1. Water in a tower has
capacity to cause change for us, if we reside at the bottom
("ground"); we could use it to cause any kind of change for us, to
some degree.

For example, we could use it to take charge -- of some limited amount -- out of the "ground" and put it on a given, heretofore
uncharged capacitor. Once the capacitor has been charged, the
charge is now at a potential above "ground." Thus, it now has
some of the capacity to cause change for us given up by the water.
If we liked, we could use the capacity now residing in the capacitor to pump water back into the tower.

How much water? Obviously not more than was used to charge
the capacitor. Obviously less; otherwise we would then have more
capacity than we had originally -- a dream. But how close could
we come to getting all the water back up? What is the theoretical
limit? Clearly that depends on-(1) how efficiently we did the
task of transferring the water's original capacity to the charge --
on what fraction of the original capacity was ultimately transferred to the charge and on what fraction was consumed -- to accomplish that transfer, and in turn, (2) how efficiently we
transfer the charge's capacity back to the water. Certainly, the
less capital we are willing to spend (on equipment and time) to

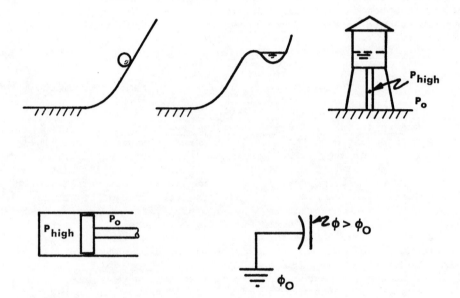

Figure 1. Examples of situations displaying a lack of complete, stable equilibrium, and hence of the potential to cause change

accomplish the two transformations, the less efficiently will we
be able to do them. Practically, whatever the desired transformation is, some capacity to cause change must be consumed by the
equipment which accomplishes the transformation. Practically, all
equipment needs to be "driven;" capacity to cause change ("fuel")
must be used up to make the equipment go.

Capital is needed to improve the efficiency of our transformations. Clearly, the worse the efficiency is in the first place,
the better the prospects for improvement. Given boundless capital
(for equipment and time) we can invest for use in charging the capacitor by lowering the water, and then for pumping back by dis-
charging the capacitor, we could come as close as we would like to
return the original amount of water to the tower, but never more.
That is the theoretical limit.

Figure 2 depicts equipment for accomplishing the transfer of
"capacity to cause change" from the charged capacitor to the
water. As the charge flows from the capacitor through the motor
its potential drops to the "ground" value -- the equilibrium value,
in our environment. The decrease in potential is given up to
torque in the drive shaft which, in turn, transmits it via the
pump to the water taken from the reservoir. The pump increases
the potential of the water, its pressure, from "ground" pressure
(atmospheric) to that pressure corresponding to the water tower
head. Thus, at the expense of capacity to cause change originally
possessed by charge on the capacitor, water with no original capacity to cause change is given such a capacity.

At an instant when current is flowing from the capacitor at
potential ϕ, and through the motor at a rate I_q, the theoretical
limit on the water flow rate I_v is given by $I_{v\,max} = (\phi - \phi_o)I_q/$
$(p - p_o)$; where ϕ_o is ground potential, p_o is "ground" (i.e.,
atmospheric) pressure at the pump inlet, and p is the pressure at
the pump outlet. The relationship for I_v follows from the fact
that the rate of hydraulic energy increase of the water $(p - p_o)I_v$
cannot exceed the rate of electric energy decrease of the charge
$(\phi - \phi_o)I_v$. The greater the "head," $(p - p_o)$, the smaller the
maximum I_v can be. Whether a small amount of water is having its
potential increased greatly or a large amount is having its potential increased slightly, the maximum "capacity to cause change"
that the water will be acquiring would be the same. That is, the
maximum $(p - p_o)I_v$ would equal the "capacity to cause change"
being given up by the charge, $(\phi - \phi_o)I_q$, which is the "potential
energy" decrease of the charge -- the energy decrease associated
with bringing it to complete equilibrium with our environment (to
"ground"). That is, under these ideal conditions with no other
energy flow besides those with I_q and I_v, the available energy
flowing out $P_{A,out} = [p - p_o]I_v$ equals the available energy flowing in with the charge which "fuels" the conversion process,
$P_{A,\,in} = [\phi - \phi_o]I_q$:

$$P_{A,out} = P_{A,in} \qquad \text{(ideal operation)}$$

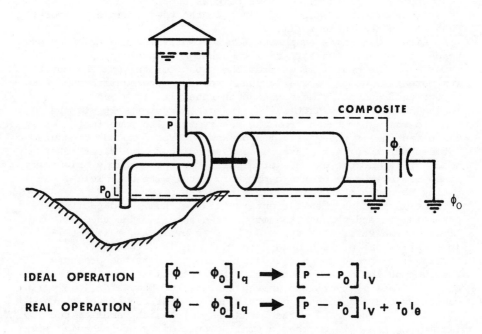

IDEAL OPERATION $\left[\phi - \phi_0\right]I_q \longrightarrow \left[P - P_0\right]I_V$

REAL OPERATION $\left[\phi - \phi_0\right]I_q \longrightarrow \left[P - P_0\right]I_V + T_0 I_\theta$

Figure 2.   Transfer of potential to cause change from one commodity (charge) to another (water)

The latter relationship would hold whether the motor is driving
the pump or whether the process were reversed and the reversed
pump (a hydraulic turbine) drove the reversed motor (an electric
generator). Likewise, it would hold if the electric motor were
replaced by a thermal motor (heat cycle) fueled by heat flowing
from a source at T greater than ambient ("ground") temperature, T_0.
And it would hold were the motor driven by a fuel (or concentration) cell, fueled by a chemical at μ_i greater than its ground
value μ_{oi}. (See (2), (3) and (4) for discussions of the ground
values μ_{oi}.) It follows that the available-energies associated
with the aforementioned currents are

$$P_A = [\phi - \phi_0]I_q$$

$$P_A = [p - p_0]I_v$$

$$P_A = [\mu_{H_2} - \mu_{0,H_2}]I_{H_2}$$

$$P_A = [T - T_0]I_\theta$$

The charge current is represented by I_q and $I_E = \phi I_q$ represents energy current. Furthermore $P_A = [\phi - \phi_0]I_q$, the current
of the commodity called available-energy, is the useful power or
available power.

Thermal Transport of Energy and Available-Energy. The energy
and available-energy currents associated with a thermal current
are $I_E = TI_\theta$ and $P_A = [T - T_0]I_\theta$. Therefore, the available-energy
current may be written in terms of the energy current as $P_A =
[1 - T_0/T]I_E$. Since the energy flow rate is the heat rate, $\dot{Q}$, it
follows that

$$P_{A,thermal} = [1 - T_0/T]\dot{Q}$$

If heat is supplied to a steady state or cyclic "heat engine"
the work output could be used to drive an electric generator for
example. If the operation (of thermal motor and electric generator) is ideal, then $P_{A,out} = P_{A,in}$. That is,

$$\dot{W}_{max} = [1 - T_0/T_{input}]\dot{Q}_{input} \qquad \text{(ideal operation)}$$

This is the classic result usually derived in a complex manner
from obtuse statements of the second law.

Potential to Cause Change for Us: A Commodity Different from
Energy. Potential energy does represent the capacity to cause
change for us. It is a commodity. It is distinct from energy; it
is not the same commodity. Energy cannot serve as a measure of
capacity to cause change for us; only potential energy (availabil-

ity) can. Some might claim the contrary, arguing that the distinction is artificial, since the difference between an energy
flow like ϕI_q with charge (or pI_v for incompressible fluids) and
the corresponding potential energy flow $(\phi - \phi_o)I_q$ is a trivial
difference which can be eliminated by measuring the potential relative to ground. Thus, $\phi_o \equiv 0$ and $\phi = (\phi - \phi_o)$. As a matter of
fact, for commodities such as charge (and volume of incompressible
fluids), which are conserved, the "ground" potential can be arbitrarily set to zero, with no disruptions. But for other, nonconserved commodities, "ground" potential cannot be set to zero; for
example, "ground" temperature T_o cannot be arbitrarily defined to
be zero.

Another important point is that the "capacity to cause
change," the potential energy, that a material has when it is not
in equilibrium with our environment in general is not simply equal
to the difference between the energy it has, E, and the energy, E_o
it would have were it brought to its "dead state," in equilibrium
with the environment. The difference between the potential energy
and $E-E_o$ stems from the fact that, while bringing the material to
equilibrium with the environment in order to get its potential
energy, it **may** be necessary to exchange things like volume and
"heat" with the environment; these exchanges will transfer energy.
Consider the confined air at $p > p_o$ and $T = T_o$ in Fig. 1. Upon
expanding, energy will be transferred to the environment to push
it aside and it will be drawn in from the environment by heat
transfer since the inside temperature tends to drop with expansion.
The net useful work output from the piston rod -- the initial potential energy of the air -- is then equal to the energy given up
by the air, $E-E_o$, plus that taken in by heat transfer $T_o(S_o - S)$
minus that given up to push aside the atmosphere, $p_o(V_o - V)$:

$$A = E-E_o + T_o(S_o - S) - p_o(V_o - V)$$

$$= E + p_oV - T_oS$$

$$- (E_o + p_oV_o - T_oS_o)$$

If the gas originally confined by the piston-and-cylinder
were not air but had a different composition, then it would not
be at completely stable equilibrium with the environment, even
when $p = p_o$ and $T = T_o$. To reduce the contents to a completely
equilibrium state, transfer of environmental components (4) to or
from the piston-and-cylinder would be necessary; thus, if an
amount $[N_{oi} - N_i]$ of component i were transferred in, it would
carry energy of amount $\mu_{oi}[N_{oi} - N_i]$. Then,

$$A = E-E_o - T_o[S - S_o] + p_o[V - V_o] - \Sigma\mu_{oi}[N_{oi} - N_i]$$

or

$$A = E + p_oV - T_oS - \Sigma\mu_{oi}N_i - [E_o + p_oV_o - T_oS - \Sigma\mu_{oi}N_{oi}]$$

The last term can readily be shown to equal zero (5). Hence
finally,

$$A = E + p_oV - T_oS - \Sigma\mu_{oi}N_i$$

This equation is an important one, for calculating the available
energy content of any material.

Also, when a flowing material is not "incompressible," but
transports available energy (and energy) that it carries as well
as that which it conveys hydraulically,

$$P_A = [e + p_ov - T_oS - \Sigma\mu_{oi}x_i]I_N + [p - p_o]I_v$$

$$P_E = eI_N + pI_v$$

Then, with $I_v = vI_N$ where v is the specific volume,

$$P_A = [e + pv - T_oS - \Sigma\mu_{oi}x_i]I_N$$

$$P_E = [e + pv]I_N$$

And, if kinetic and gravitational energy are negligible, e + pv =
h, the so-called enthalpy.

<u>Available Energy Consumption.</u> In contrast with energy and
charge, available-energy is not a conserved commodity. Available-
energy is called "energy" in lay terminology, and is the true
measure of the potential of a substance to cause change; some is
destroyed (consumed) in any real process. The unreal, ideal operation referred to obove when $P_{A,out} = P_{A,in}$, is the theoretical
limit which can be approached, but never reached in practice.
Associated with real motors and pumps, there will always be dissipations óf potential energy -- consumption thereof--used up to
make the motor and pump "go." These dissipations manifest themselves in "heat production;" if steady operating conditions are to
be maintained -- which we will assume here, since it will help illustrate certain important points -- the "heat" (entropy) which
is produced must be transferred away, eventually flowing into our
atmosphere at "ground" temperature, T_o. The thermal current into
the atmosphere, I , will need to equal the rate of "heat" (entropy) production in this steady case, and the associated energy
transfer will be T_oI_θ. The energy balance for the composite,
saying energy efflux equals energy influx, now yields $\phi I_q + p_oI_v =$
$\phi_oI_q + pI_v + T_oI_\theta$. Hence, $(p - p_o)I_v = (\phi - \phi_o)I_q - T_oI_\theta$. That
is, the potential energy output will be less than the input by the
amount consumed (used up, destroyed, annihilated) to "drive" the

transformation:

$$P_{A\ out} = P_{A\ in} - \dot{A}_c$$

where $\dot{A}_c = T_o I_\theta$ represents the rate of available energy consumption -- rate of potential energy consumption.

The thermal current I_θ leaving the composite in Figure 2 is the rate at which "heat" is being produced inside the composite. It can be readily shown that for any system (4)

$$\dot{A}_c = T_o \dot{S}_p$$

where $\dot{S}_p$ is the rate of entropy production within the system.

The Second Law. In summary, then, energy does not, in general, represent the "capacity to cause change for us;" energy flows associated with nonconserved commodities are not representative of such capacity. And, energy associated with such commodities cannot, even in the ideal limit, be completely transferred to other commodities.

Potential energy, which anything has when it is not in complete equilibrium with our environment, does represent the capacity to cause change for us; it can be transferred from one thing to any other, completely in the ideal limit. In actuality, to accomplish changes for us some potential energy is invariably used up, because it is needed to make the changes occur. (Therein lies its value!) This paragraph presents the essence of the Second Law.

Energy is not the commodity we value; potential energy (availability) is.

The Roles of Thermodynamics

Traditionally, Thermodynamics has served the following purposes:

1. It provided the concept of an energy balance, which has commonly been employed (as one of the "governing equations" (9)) in the mathematical modelling of phenomena. (However, it is not necessary to use an energy balance. An entropy (or available energy) balance can be used instead (9).)
2. It has provided mathematical formulas for evaluating properties such as enthalpy and entropy from property relations, determined by direct or indirect experiment, and from partial derivatives of the property relations.
3. It has provided the means for establishing the final equilibrium state of a system in any given initial state and subjected to given constraints.

Now, with more modern formulations of Thermodynamics, it can be used for the following purposes as well:

4. Pinpointing the inefficiencies in and losses from processes, devices and systems. The concept of available energy is needed for this purpose; attempts to use energy for gauging efficiency leads to erroneous results -- often grossly erroneous.

5. Cost accounting of "utilities"; that is, of "energy" services. This is useful in engineering (design; operation of systems), and in management (pricing; calculating profits). Again, the key is the use of available energy, and not energy (6,7).

6. The governing equations for a phenomenon can be derived, by selecting the appropriate commodity balances (those for all commodities transported and/or produced during the phenomenon) and utilizing the First and Second Laws (8,9). The derivations can also be accomplished for highly nonequilibrium processes (9), by basing the Second Law on available energy (and replacing the concept called reversibility by the more general concept, identity).

The roles of primary interest in this volume are those related to the direct practical application of available energy.

Literature Cited

1. Obert, E.F., *Elements of Thermodynamics and Heat Transfer,* McGraw-Hill, 1949.

2. Gaggioli, R.A. and Petit, P.J., "Use the Second Law First", *Chemtech,* pp. 496-506, August, 1977.

3. Rodriguez, L., "Calculation of Available Energy Quantities", this volume.

4. Wepfer, W.J. and Gaggioli, R.A. " Reference Datums for Available Energy," this volume.

5. Obert, E.F., *Concepts of Thermodynamics,* McGraw-Hill, 1960, See Equation 14-26.

6. Reistad, G.M. and Gaggioli, R.A., "Available Energy Accounting", this volume.

7. Wepfer, W.J., "Applications of Available Energy Accounting", this volume.

8. DeGroot, S.R. and Mazur, P., *Nonequilibrium Thermodynamics,* North-Holland, Amsterdam, 1962.

9. Gaggioli, R.A. and Scholten, W.B., "A Thermodynamic Theory for Nonequilibrium Processes", this volume.

RECEIVED October 17, 1979.

Second Law Procedures for Evaluating Processes

PETER J. PETIT

Coal Gasification Systems Operation, Allis–Chalmers Corporation,
Milwaukee, WI 53201

RICHARD A. GAGGIOLI

Department of Mechanical Engineering, Marquette University,
1515 W. Wisconsin Ave., Milwaukee, WI 53233

Josiah Willard Gibbs and James Clerk Maxwell gave form to the concept of "available energy" more than one hundred years ago; however, efforts in this century to popularize its use in evaluating energy conversion processes have met with limited acceptance.

Available energy is a property which measures a substance's maximum capacity to cause change, a capacity which exists because the substance is not in equilibrium with the environment. It is any form of potential energy. ("Potential energy" is used here in a broader sense than the traditional concept of energy associated with a conservative force field.) Consequently, it is a perfectly rational basis for assigning value to a fuel--whether that fuel be coal, steam, electricity, water in an elevated reservoir, or any other commodity having the potential to drive a process. Available energy is destroyed, and the fuel utilization efficiency reduced, in any process where a potential (voltage, pressure, chemical, thermal, etc.) is allowed to decrease without causing a fully equivalent rise in some potential elsewhere. It is a simple and understandable concept, completely consistent with our intuition and everyday perceptions; it is what the layman calls "energy".

Unfortunately, another property, called energy by scientists and engineers, has become the traditional basis for assigning fuel value to substances. And because of this, process efficiencies have come to be defined as energy ratios. Energy efficiency is only an approximation of the true efficiency with which a fuel resource is used, and often a poor one. Why have designers of energy conversion systems settled for this approximation for such a long time? What has prevented available energy from taking its rightful place as the true yardstick of fuel value and process efficiency?

One reason has had its foundation in economics. Fuels were so abundant and cheap that the fuel contribution to product cost was small relative to the contribution of plant capital cost. Consequently, a more handsome payoff could usually be achieved by focusing engineering resources on lowering capital costs instead of improving fuel efficiency. With little incentive to improve

0–8412–0541–8/80/47–122–015$05.75/0
© 1980 American Chemical Society

efficiency, there has been little motivation to understand and apply available energy.

A second barrier has been the slow historical development of available energy theory, due in part to the lack of economic incentive but also a consequence of the fact that scientific progress is typically evolutionary. Furthermore, it has been a common viewpoint until quite recently that the development of Thermodynamics as a subject was virtually complete, and that little further investment of scientific research was warranted. It is quite clear now that this is not the case. Thermodynamic theory is receiving renewed interest, and deservedly so for many reasons. (Though open to controversy, it is probable that some of the apparently most profound developments in the history of Thermodynamics have detoured its advance and/or been improperly understood.)

Today, however, with energy costs rising, the importance of second law procedures is becoming recognized, and they will become widespread. This, along with scientific curiosity, has provided a major incentive for lowering the theoretical barriers as well. Thermodynamics can now be presented in a more straightforward manner.

The Role of Second Law Analysis

Available energy analysis is intended to complement, not to replace, energy analysis. Energy balances, when used in conjunction with mass balances and other theoretical relations, help produce a workable design (Figure 1). Those workable designs that satisfy other imposed criteria (such as plant capacity) and constraints (such as environmental standards) can be called acceptable. But of all the conceivable acceptable designs, there is generally only one design for which the product unit cost is minimized--the optimal design. The principal role of available energy analysis is to assist in approaching the optimal design.

One of two ways in which available energy analysis assists is by pinpointing and quantifying both the consumptions of available energy, used to drive processes, and effluent losses of available energy. These are the true inefficiencies, and therefore they point the way to improvement of a design. Furthermore, for the same reason the second law analysis stimulates creativity leading to entirely new concepts--new technology.

Another manner in which available energy can be employed for design optimization is with available energy accounting (1). Available energy, being the true measure of the fuel value of any commodity, provides a common and rational basis for costing all the flow streams, heat transports and work transfers in an energy-conversion or chemical-process system. Consequently, the traditional tradeoff between the operating and capital costs can be optimized unit by unit within the system. (Available energy costing is of value not only for design optimization, but also for cost accounting purposes. Methods of costing available energy

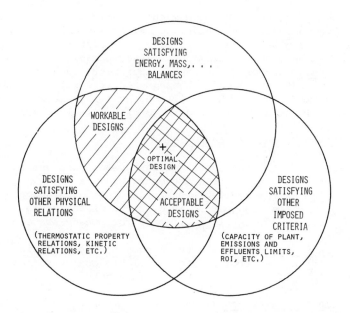

Figure 1.   Roles of energy balances in the creation of acceptable designs

are treated elsewhere ($\underline{1}$,$\underline{2}$,$\underline{3}$,$\underline{4}$).)

In the role of design optimization it is available energy analysis, not energy analysis, which is the appropriate tool, because available energy is the "common denominator." That is, all forms of available energy are equivalent to each other with respect to their capacity to cause change. The same cannot be said for all forms of energy. For example, shaft work can be turned entirely into electricity (theoretically, at least) while the thermal energy in steam cannot. However, the available energy content of that steam could be turned entirely into electricity (or entirely into shaft work).

To defend process design decisions or support energy policy arguments, it is common to hear qualitative arguments like "high grade energy such as electricity is worth more than lower grade energy such as steam." But how much electricity is that steam worth? Available energy analysis provides the means to present that kind of argument quantitatively. The electrical equivalent of steam can be simply determined. (And, the determination can be made without dragging extraneous equipment, such as a turbine, into the evaluation, as would be required if first law techniques were to be used. Those who insist on involving extraneous equipment are failing to grasp the essence of available energy.)

Tools Used in Second Law Analysis

Available energy analyses and energy analyses use the same family of tools to evaluate and compare processes:

1) balances for available energy and for each independent commodity which is transported into or out of the system (such as $\Sigma \dot{E}_{transports} = 0$, $\Sigma \dot{A}_{transports} = \dot{A}_{\delta}$ for steady-state operation),

2) transport relations between companion commodities (such as that relating the flow rate of available energy to that of entropy: $\dot{A}_S = (T - T_0)\dot{I}_S$; or that relating the flow rates of energy and volume: $\dot{E}_V = p\dot{I}_V$),

3) kinetic relations (like $\dot{Q} = UA\Delta T_m$, or $r_{AB} = k_{AB}c_A{}^a c_B{}^b$), which relate transports or productions (reaction rates) to driving forces, and

4) thermostatic property relations specific to the material involved (such as s(T,p) and h(T,p) for H_2O as embodied in the Steam Tables, or $s_2 - s_1 = c_p \ln T_2/T_1 - R \ln p_2/p_1$ for perfect gases, or the absolute entropy of CO_2 at 250°C and 1 atm, or the standard enthalpy of formation of CH_4, etc.).

Of these four tools, only the first two, balances and transport relations, need further discussion here, inasmuch as they differ for available energy analyses.

Available Energy Balances. Writing a steady-state balance for available energy is just like writing a steady-state energy balance except for one major difference. While energy is conserved, available energy can be destroyed (not lost, but actually consumed), and so the balance must contain a destruction term:

Total Available Energy Transported into the System	=	Total Available Energy Transported out of the System	+	Available Energy Destroyed within the System
$\Sigma \dot{A}_{in}$	=	$\Sigma \dot{A}_{out}$	+	$\dot{A}_{\delta}$

When the transport rates of independent commodities are known (given or determined from kinetic relations), then the available energy transport terms can be evaluated using the aforementioned relations. (The application of these transport relations, which will now be set forth, requires the use of the thermostatic property relations.) Once the transport term values are known, the balance can be used to evaluate the available energy destruction, $\dot{A}_{\delta}$.

Transport Relationships. The following expressions are used to evaluate the transports, $\dot{A}_{in}$ and $\dot{A}_{out}$, of available energy.

a) Shaft Work: When energy and available energy are transported via a turning shaft--with torque, τ, which is simply a current, I_{α}, of angular momentum--the energy flow rate is $\dot{E} = \omega \cdot I_{\alpha}$ where ω is the angular velocity. This relation is usually written as $\dot{W} = \omega \cdot \tau$ since the energy flow rate, $\dot{E}$, is the same as the work rate, $\dot{W}$, and the flow rate of angular momentum, I_{α}, is the torque, τ.

The available energy current is given by

$$\dot{A} = [\omega - \omega_0] \cdot I_{\alpha} = [\omega - \omega_0] \cdot \tau$$

Since the angular velocity ω_0 of the environment is zero (for phenomena for which the earth can be taken as an inertial reference frame),

$$\dot{A}_{shaft} = \omega \cdot \tau$$

Which is identical to the work rate, $\dot{W}$

$$\dot{A}_{shaft} = \dot{W}_{shaft}$$

As a consequence, the conclusion can be drawn, from the second law, that "the available energy is the maximum shaft work obtainable." This statement is usually used to <u>define</u> available energy. Unfortunately such a definition gives the impressions (i) that available energy is relevant only to "work processes," and (ii) that work is the ultimate commodity of value. Actually, available energy is the commodity of value, regardless of the form (thermal, work, chemical, electrical, . . .); and it is relevant to processes involving any of these forms.

b) Thermal Transports of Available Energy: The energy and available energy currents associated with a thermal current at a temperature T are $\dot{E}_S = TI_S$ and $\dot{A}_S = (T - T_0)I_S$ (5). By combining these two expressions, the available energy current can be written in terms of energy current as $\dot{A}_S = (1 - T_0/T)\dot{E}_S$. Since the energy flow rate by heat transfer is usually represented by $\dot{Q}$,

$$\dot{A}_{thermal} = (1 - T_0/T_Q)\dot{Q}$$

If $\dot{Q}$ represents the energy supplied at a temperature T_Q to a steady-state or cyclic "heat engine" (Figure 2), it follows from an available energy balance that the net rate of available energy flowing from the cycle in the form of shaft work can at most be equal to the thermal available energy supplied to the cycle; i.e.,

$$\dot{A}_{shaft} \leq \dot{A}_{thermal}$$

Using the transport relationship $\dot{A}_{shaft} = \dot{W}_{shaft}$, we may write

$$\dot{W}_{max} = (1 - T_0/T_Q)\dot{Q}$$

This is the classic result usually derived in a complex manner from traditional (obtuse) statements of the second law. (When the heat transport occurs over a range of temperatures, as in a conventional steam cycle, then the integral form of these equations must be used: $\dot{W}_{max} = \dot{A}_{thermal} = \int (1 - T_0/T_Q)\,d\dot{Q}$.)

c) Simultaneous Thermal and Chemical Available Energy Flows with Matter: The energy and available energy flows associated with transports of material j are:

THERMAL: $\dot{E}_S = TI_S$ and $\dot{A}_S = [T-T_0]I_S$

CHEMICAL: $\dot{E}_j = \mu_j I_j$ and $\dot{A}_j = [\mu_j - \mu_{j0}]I_j$

where μ_{j0} is the reference chemical potential of material j in the environment. The energy current for simultaneous thermal and chemical transfers associated with the flow of material j is

$$\dot{E}_j = \mu_j I_j + TI_S = (\mu_j + Ts_j)I_j \qquad (\text{since } I_S = s_j I_j)$$

$$= h_j I_j$$

The available energy flow is $\dot{A}_j = (\mu_j - \mu_{j0}) I_j + (T - T_0)s_j I_j$, which reduces to

$$\dot{A}_j = (h_j - T_0 s - \mu_{j0})I_j$$

Evaluation of Available Energy Transport Expressions.
Available energy transport relations are seen to be products of
thermostatic properties with commodity currents. Given the commodity currents, the available energy transports can then be evaluated by determining the thermostatic properties, using traditional thermochemical property evaluation techniques. References (6)
and (7) present convenient relationships for practical evaluation
of available energy flows for several important cases.

A prerequisite for the evaluation of the available energy
transports is the selection of a proper reference environment.

The amount of available energy which a substance has is relative and depends upon the choice of a dead state. The fundamental dead state is the state that would be attained if each constituent of the substance were reduced to complete stable equilibrium
with the components (8,9,10) in the environment--a component-equi-
librium dead state. (Thus, one may visualize the available energy
as the maximum net work obtainable upon allowing the constituents
to come to complete equilibrium with the environment.) The equilibrium is dictated by the dead state temperature T_0; and, for
ideal gas components, by the dead state partial pressure p_{j0} of
each component j. (The available energy could be completely obtained, say in the form of shaft work, if equilibrium were
reached via an ideal process--no dissipations or losses--involving
such artifices as perfectly-selective semi-permeable membranes,
reversible expanders, etc. (9,10,11).)

Second Law Efficiency--The True Efficiency

In the theoretical limit, available energy contained in any
commodity can be completely transferred to any other commodity
(12,13). In the case of real transformations, the degree to which
this perfection is approached is measured by the second law efficiency (often called the "effectiveness" (8,9,10,11):

$$\eta_{II} = \frac{\text{available energy in useful products}}{\text{available energy supplied in "fuels"}}$$

The denominator exceeds the numerator by the amount of available
energy consumed by the transformation plus the amount lost in effluents:

$$\eta_{II} = \frac{\dot{A}_{products}}{\dot{A}_{products} + \dot{A}_{destroyed} + \dot{A}_{lost}}$$

For any conversion, the theoretical upper limit of η_{II} is 100%,
which corresponds to the ideal case with no dissipations. To
approach this limit in practice requires the investment of

greater and greater capital and/or time. The tradeoff, then, is
the classical one: operating costs (for fuel) versus capital (for
equipment and time). The important point here is that attainment
of the optimal design can be greatly facilitated by application of
Second-Law analyses (i.e., available energy analyses) to processes,
devices, and systems.

Traditional efficiencies (here called first law efficiencies,
η_I) based on the ratio of "product" energy to "fuel" energy are
generally faulty, to a degree that depends on the kind of device
or system to which they are applied. Basically, the worth of a
first-law efficiency is proportional to how well it approximates
the second law efficiency.

Because energy is conserved, the difference between the energy output in the products from a system and the energy input with
fuels--the difference which, it is supposed, represents the inef-
ficiency--must be the energy lost with effluents. Available
energy ("fuel") consumptions, which drive the various operations
in the system, are neglected. Thus, electric power plants; which
are among the truly most efficient ($\sim$ 35%) energy-conversion systems appear to be very inefficient by the usual energy standards,
because of large energy losses. Whereas, the usual comfort-heat-
ing furnace, one of the most inefficient (η_{II} = 7 - 12%) energy
converters, seems very efficient (η_I = 70%). The notorious "heat
losses" from power plants are hardly losses at all, accounting
for only 5% of the fuel input; the major inefficiencies (45%)are
within the boiler, which appears to be very efficient in terms
of energy (η_I = 90%).

The Methodology of Available Energy Analyses

How the tools are organized into a methodology for process
evaluation via available energy will be illustrated in this paper
with the help of a very simplified coal-fired boiler, often found
in textbooks on thermodynamics. It will be used to demonstrate
the calculation of available energy flows, losses and consumptions.

Application to Coal-Fired Boiler. Consider this problem: A
given coal-fired boiler is burning Illinois No. 6 coal while converting 298°K (77°F) water to 755°K (900°F) steam, at 5.86 MPa
(850 psia). The performance of the boiler under these conditions
is reflected by a first law efficiency (η_I) of 85%. How much of
the coal's available energy is destroyed? What is the second law
efficiency (η_{II}) of the boiler? Where are the distinct available
energy consumptions within the boiler and what are their magnitudes? Where are the losses, and what are their quantities? Figure 3 illustrates one type of flow diagram which can be drawn for
this boiler. As in energy analyses, it serves to define the boun-
daries of the process being studied as well as to establish a system for stream identification. For convenience, key stream

properties have been included in Figure 3.   The characteristics of
the coal are presented in Table I.   (To retain perspective in this
example, all flows will be placed on a "per unit weight of coal
input" basis.)

TABLE I
PROPERTIES OF ILLINOIS NO. 6

Ultimate Analysis:

C	.606
H	.054
O	.091
N	.014
S	.035
M	.100
ASH	.100

HHV = 25607 kJ/kg raw coal = 11011 Btu/lb raw coal
LHV = 24429 kJ/kg raw coal = 10504 Btu/lb raw coal

With the given first law efficiency, the energy supplied to
the H_2O is 0.85(25607) = 21766 kJ/kg coal = 9359 Btu/lb coal.
The property relationship between temperature, pressure and enthalpy for H_2O permit finding the flow rate of steam:

$$\dot{m}_S = \frac{\Delta H}{[h(T_S, P_S) - h(T_F, P_F)]}$$

$$= \frac{21766}{(3381 - 110.6)} = 6.6555 \; \frac{kg \; steam}{kg \; coal}$$

An available energy balance on the same system says

$$\Sigma \dot{A}_{in} \qquad = \qquad \Sigma \dot{A}_{out} \qquad \dot{A}_\delta$$

$$[\dot{A}_C + \dot{A}_A + \dot{A}_F] = [\dot{A}_S + \dot{A}_{L'} + \dot{A}_{G'}] + \dot{A}_\delta$$

To determine the consumption term, $\dot{A}_\delta$, the transport terms will be
evaluated first.

In the boiler problem, the stable reference environment
(Table II) is taken as atmospheric air saturated with H_2O at
$T_0 = 298.15^\circ K$ ($77^\circ F$) and $p_0 = 101320$ Pascals (14.696 psia) in
equilibrium with several condensed phases ($\underline{12}$).   These values of
T_0 and p_0 were chosen because they are representative of yearly
average air conditions for much of the contiguous United States
and because no adjustments for temperature need to be made when
applying tables of heats of formation, absolute entropies and
free energies of formation.   Saturated air was chosen so that
liquid water at T_0, p_0 would be in complete equilibrium with the

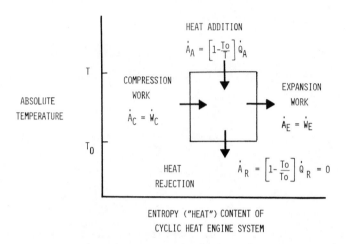

Figure 2.   *Transports of energy and available energy in a thermodynamic cycle*

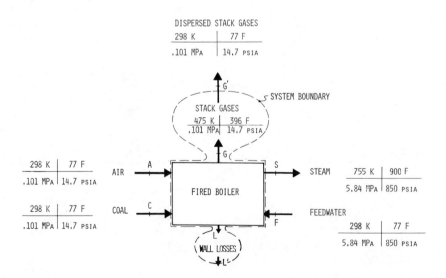

Figure 3.   *Flow diagram of fired boiler (with stream labels and key property data)*

TABLE II
CHARACTERISTICS OF STABLE REFERENCE ENVIRONMENT

T_0 = 298.15°K (77°F) P_0 = 1 atm

Composition:

Air Constituents	Mole Fraction
N_2	.7560
O_2	.2034
H_2O	.0312
A	.0091
CO_2	.0003
H_2	.0001

Condensed phases at T_0, P_0: H_2O; $CaCO_3$, $CaSO_4 \cdot 2H_2O$

air, consistent with the concept of a <u>stable</u> reference environment.

If a lower-temperature sink were involved in the problem (such as cooling water for a power plant from a lake) then it would be appropriate to use the temperature of that sink as T_0.

<u>Evaluation of Transports</u>. The available energy transported with the <u>coal</u> may be estimated using the appropriate formulas of Szargut and Styrylska; see (<u>6</u>). Noting that the mass ratio of oxygen to carbon is 0.15,

$$\beta_1 = 1.0438 + 0.1882 \frac{H}{C} + 0.0610 \frac{O}{C} + 0.0404 \frac{N}{C}$$

$$= 1.07066, \text{ dimensionless ratio of } a_c \text{ to LHV.}$$

$\dot{A}_c$ = available energy transport with coal

$$= \beta_1 \cdot \text{LHV} + 6740S = 1.07066(24429) + 6740(0.035)$$

$$= 26391 \frac{kJ}{kg \text{ raw coal}} = 11348 \text{ Btu/lb raw coal}$$

Notice that, as required in (<u>6</u>), the LHV of dried fuel per unit of moist fuel is used here. This simply means that in the calculation of LHV from HHV, the <u>water formed during combustion</u> is assumed to be <u>vapor</u> in the final state, but the <u>water originally present</u> as moisture in the coal is assumed not to have been vaporized by the combustion process. In this problem, therefore,

$$LHV = HHV - H \frac{18.016}{2.016} (h_{fg} @ T_0)$$

$$= 25607 - 0.054 \frac{18.016}{2.016} (2442)$$

$$= 24427 \frac{kJ}{kg \text{ moist fuel}}$$

If the LHV is calculated assuming that the original moisture in the coal as well as that formed during combustion are in the vapor state (not the usual procedure), it can be corrected for the latent heat of coal moisture as shown in (6):

$$LHV = LHV_{wet} + 2442\dot{M}$$

The combustion air, free from the environment, has zero availability:

$$\dot{A}_A = \text{available energy transport with combustion air} = 0.$$

The transport of available energy with the feedwater is due only to its pressure since it is at T_0 and water is "free" from the environment (except for purification). $\dot{A}_F$ is equal to the ideal pump work required to impart that pressure to the water. Assuming the water is incompressible (6):

$$\dot{A}_F = \dot{m}_F \bar{v}(p_F - p_0) = 50.2 \frac{kJ}{kg \text{ coal}} = 21.6 \text{ Btu/lb coal}$$

To evaluate the transport of available energy with steam, the dead state values of enthalpy and entropy for H_2O must be determined (note that H_2O is liquid at T_0 and p_0):

$$h_0 = h_f(T_0) + [h_0 - h_f(T_0)]$$

$$= h_f(T_0) + v_f[p_0 - p_{sat}(T_0)] \doteq h_f(T_0)$$

$$= 104.84 \frac{kJ}{kg} = 45.083 \text{ Btu/lb}$$

$$s_0 = s_f(T_0) \qquad \text{(Entropy varies little with pressure in this region)}$$

$$= 0.3671 \frac{kJ}{kg°K} = 0.0877 \text{ Btu/lb°R}$$

Now the available energy in the steam can be easily evaluated using property relations from the steam tables:

$\dot{A}_S$ = available energy transport with steam

$$= \dot{m}_S \left[(h - T_0 s) - (h_0 - T_0 s_0) \right]$$

$$= 6.6555 \left[(3381 - 298.15(6.8383)) - (104.84 - 298.15)(0.367)) \right]$$

$$= 8963.3 \frac{kJ}{kg\ coal} = 3854.2\ Btu/lb\ coal$$

For convenience, the system boundary has been located far from the stack, so that the stack gases have (virtually) been fully dispersed. Thus, the available energy of the <u>dispersed stack gases</u> leaving the system is practically zero;

$\dot{A}_{G'}$ = available energy of dispersed stack gases = 0.

This artifice avoids the calculation of the available energy <u>loss</u> at the stack exit. Then, the total available energy destruction with the system will include that loss--inasmuch as the lost available energy is ultimately consumed (destroyed) by the dispersion process.

Similarly, the available energy lost by heat transfer from the outer surface of the boiler to the ambient is eventually consumed, as the heat approaches T_0. Thus, by locating the system boundary so that the heat flow across it is at T_0,

$\dot{A}_{L'}$ = available energy escaping with heat transport to

$$\text{the environment} = [1 - \frac{T_0}{T_0}]\dot{Q}_{L'} = 0$$

Now that all the transports of available energy across the system boundary have been evaluated, the <u>available energy consumption</u> term can be determined:

$$\dot{A} = \dot{A}_{in} - \dot{A}_{out} = \dot{A}_C + \dot{A}_A + \dot{A}_F - \dot{A}_S - \dot{A}_{G'} - \dot{A}_L$$

$$= 26391 + 0 + 50.2 - 8963.3 - 0$$

$$= 17478 \frac{kJ}{kg\ coal} = 7515\ Btu/lb\ coal$$

In answer to the second question, the <u>second law efficiency</u> of this system is

$$\eta_{II} = \frac{A_S - A_F}{A_C} = 0.338$$

compared to the first law efficiency of

$$\eta_I = \frac{H_S - H_F}{H_C} = 0.85$$

Analysis of Sub-processes. To determine the locations and magnitudes of the consumptions which comprise $\dot{A}_\delta$, one need only subdivide the system appropriately into subsystems, and then repeat the foregoing procedure. Thus, the boiler in this problem can be broken down into three separate processes: 1) combustion, 2) heat transfer, and 3) dissipation of the stack gases. Each can be analyzed for its second law efficiency and the amount of available energy it consumes.

To analyze the last of these sub-processes, let us return to the stack gas transport of available energy which was side-stepped before. The system boundary, represented by the dashed line in Figure 3, can be redrawn so that the stack gases cross it just as they leave the stack, at $T_G = 475°K = 396°F$. At this location (G) the gases are at the same pressure as the environment but are not in thermal or chemical equilibrium with it. Even if cooled at T_0 the stack gases at a total pressure p_0 would still not be in complete equilibrium with the environment, because the composition is different. The stack gas properties are given in Table III. These properties have been determined, using property relations, by applying mass and energy balances to the boiler.

TABLE III
PROPERTIES OF STACK GASES IN COAL-FIRED BOILER PROBLEM

Gas Component	Mass Rate Per Unit Mass of Coal	Mole Fraction
CO_2	2.2205	.14517
H_2O	.5826	.09305
SO_2	.0699	.00314
O_2	.1987	.01787
N_2	7.2127	.74077
ASH	.1000	--
TOTAL	10.3844	1.00000

$$c_p(T) \doteq 0.9385 + 0.3118(T/10^3) + 0.0328(T^2/10^6)$$
$$- 0.0383(T^3/10^9) \quad [kJ/(kg°K)]$$

$$T_G = 475°K \ (396°F) \qquad p_G = .101 \ MPa \ (14.7 \ psia)$$

$$AMW = 29.592 \ g/gmole$$

Using formulas presented in (6), the total available energy in the stack gases may be calculated as the sum of three contributions, which for convenience (but without attaching strict physical significance) can be called thermal, pressure, and chemical. The available energy contributions in the <u>stack gas</u> can be determined as follows:

Thermal:

$$\dot{A}_{therm} = \dot{m}_G \int_{T_0}^{T_G} c_p \left(1 - \frac{T_0}{T}\right) dT = 423.7 \frac{kJ}{kg\ coal}$$

$$= 182.2\ Btu/lb\ coal$$

Pressure:

$$\dot{A}_{press} = \dot{m}R_0 T_0 \ln(p_G/p_0) = 0 \quad since\ p_G = p_0$$

Chemical:

$$\dot{A}_{CO_2} = \dot{m}_{CO_2} (2.4789\ \ln 0.14517 + 20.108)(1000)/44.011$$

$$= 773.2 \frac{kJ}{kg\ coal}$$

$$\dot{A}_{H_2O} = 87.6 \qquad \dot{A}_{SO_2} = 220.9 \qquad \dot{A}_{O_2} = -37.4$$

$$\dot{A}_{N_2} = -13.1 \qquad\qquad \dot{A}_{ASH} = 0$$

$$\dot{A}_{chem} = \Sigma A_i = 1031.2 \frac{kJ}{kg\ coal} = 443.4\ Btu/lb\ coal$$

Total:

$$\dot{A}_G = available\ energy\ loss,\ transported\ with\ stack\ gases$$
$$(and\ eventually\ destroyed)$$

$$= \dot{A}_{therm} + \dot{A}_{press} + \dot{A}_{chem} = 1454.9 \frac{kJ}{kg\ coal}$$

$$= 625.6\ Btu/lb\ coal$$

To calculate the consumptions of available energy in the combustion process and the heat transfer process, it is supposed that the boiler may be separated into two distinct entities (Figures 4 and 5). The transports of available energy into the combustion process with air, steam, feedwater and stack gases have already been determined. Assuming that the <u>products of combustion</u> have the same composition and total pressure as the stack gases, the

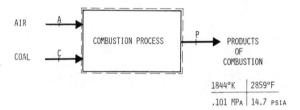

Figure 4.   Flow diagram of combustion process in coal-fired boiler problem

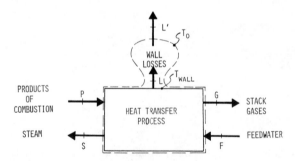

Figure 5.   Flow diagram of heat transfer process in coal-fired boiler problem

chemical and pressure contributions to total available energy need
not be redetermined. Once the temperature T_P of the combustion
products is determined (via an energy balance and thermochemical
property relations), the thermal contribution may be evaluated as
follows:

$$\dot{A}_{therm} = \dot{m}_P \int_{T_0}^{T_P} c_p (1 - \frac{T_0}{T}) dT = 18458.1 \frac{kJ}{kg\ coal}$$

$$= 7937.0\ Btu/lb\ coal$$

$\dot{A}_p$ = available energy transport with products of combustion

$$= \dot{A}_{therm} + \dot{A}_{press} + \dot{A}_{chem} = 19489.3 \frac{kJ}{kg\ coal}$$

$$= 8380.4\ Btu/lb\ coal$$

Now, having evaluated the relevant transport terms, two important
consumptions of available energy within the boiler may be evaluated, by applying an available energy balance to the "combustor"
and one to the "heat exchanger":

$\dot{A}_{\delta,RXN}$ = destruction of available energy due to uncontrolled
combustion of coal

$$= \dot{A}_C + \dot{A}_A - \dot{A}_p = 6901.7 \frac{kJ}{kg\ coal} = 2967.7\ Btu/lb\ coal$$

$\dot{A}_{\delta,HT}$ = destruction of available energy due to the heat
transfer process

$$= \dot{A}_P - \dot{A}_G + \dot{A}_F - \dot{A}_S = 9121.3 \frac{kJ}{kg\ coal}$$

$$= 3922.2\ Btu/lb\ coal$$

Correspondingly, the second law efficiency for each of these internal processes may now be evaluated:

$$\eta_{II,RXN} = \frac{\dot{A}_p}{\dot{A}_C} = 0.738$$

$$\eta_{II,HT} = \frac{\dot{A}_S - \dot{A}_F}{\dot{A}_p - \dot{A}_G} = 0.494$$

Figure 6 shows one method of presenting the results of an
available energy analysis. It is similar to an energy flow diagram, with the added feature of showing consumptions of available
energy as negative values <u>within</u> the various process blocks. Such
a diagram aids in gauging the relative importance of each

transport and consumption. (The so-called Sankey diagram has been
used effectively in the European literature, representing each
available energy flow by a band so drawn that its width is proportional to the flow.)

Discussion. The "thermal efficiency" of this boiler--that is,
the net usable heat output in steam divided by total heat input in
coal--is purported to be 85%. To cite such efficiencies--energy
ratios--is misleading. As shown by the foregoing overall analysis
of the boiler, the "available energy" of the steam, its useful
energy, is much less than its energy content; hence, the energy
efficiency, η_I, cited for the boiler is 2-1/2 times its true efficiency, η_{II}, of 33.8%.

The detailed analysis of the different sub-processes of the
boiler, as summarized in Figure 6, shows that the two largest dissipations are due to the uncontrolled kinetics of combustion
(26.2% of the total available energy input), and heat transfer
(34.5%) lost as heat passes from hot products at a high average
$(1 - T_0/T)$ to liquid and gaseous H_2O with a relatively low average
$(1 - T_0/T)$. The stack losses, while not insignificant, represent
only 5.5% of the available energy in the coal; in contrast, they
represent nearly 15% of the coal's energy content.

Of course, no cost effective opportunities to reduce any
consumption or loss should be overlooked. Often, the better opportunities are where the larger consumptions (and losses) occur.
For example, if the steam pressure were raised, the average temperature and, therefore, the average value of $(1 - T_0/T)$ for heat
addition to the steam would be raised and a significant decrease
in $\dot{A}_{\delta,HT}$ could be affected. (Design modifications may have to be
made in the equipment utilizing the steam as "fuel," in order to
effectively take advantage of the steam's higher available energy
content. That is, improving η_{II} of the boiler does not necessarily imply an improvement in the overall η_{II} for the overall process
of which the boiler is only a part. If the requirements for steam
are at low-pressure, it would be of no benefit to produce it at
high pressure and then simply throttle it to the needed pressure.
On the other hand, if a turbogenerator were used in lieu of a
throttling valve, electric power could be obtained while dropping
the steam to the desired pressure--cogeneration.) In effect, the
steam's available energy content would be higher, without increasing the available energy input (coal) to the boiler.

A detailed available energy analysis was carried out in reference (13) on a modern 300-MW coal-burning power plant. The
available energy flows calculated in that analysis are presented
in Figure 7. Corresponding energy flows are included in parentheses for comparison.

Also notable is that half of the potential energy of the incoming fuel is destroyed immediately by the boiler (30% is used
up in combustion, 15% is consumed in the heat transfer from high-
temperature products to the steam and preheated air, and 5% is

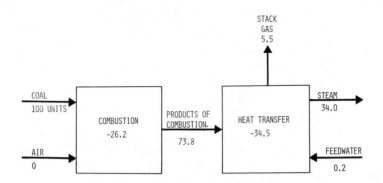

Figure 6.   Available energy flow diagram for coal-fired boiler problem

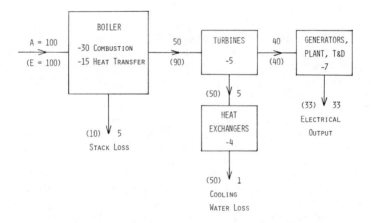

Figure 7.   Available energy (and energy) flow diagram for a conventional fossil-fired steam power plant (negative numbers are available energy consumptions)

lost with the stack gases).

The corresponding energy balance implies that the boiler is very efficient, losing only 10% of the input energy--virtually all of it associated with stack gas thermal losses.

"Energy losses" associated with the condenser, carried into the environment by the cooling water, are great. We hear much about the need to utilize that energy. Actually, virtually none (< 2%) of the resource which went into the power plant is lost in that water. The real loss was (primarily) back in the boiler where "heat" (entropy) was produced. Once produced, it must be transmitted to "ground"--the environment--to obtain its potential energy. It carries much energy with it to the environment but essentially no potential energy. Attempts to take advantage of "all that energy" and thereby improve the utilization of the fuel used by the plant are futile. On the other hand, the renewed interest in the cogeneration of electricity and "heat" is consistent with Second Law results. As described earlier, cogeneration reduces the boiler inefficiencies, associated with the production of steam (or hot water) at low temperatures. (Notice that the production of hot water by cogeneration is tantamount to raising the turbine back-pressure, so that the condensing steam, at a high temperature, can raise the temperature of the cooling water more. Then, the cooling water becomes useful, because it has substantial available energy. Of course, electrical available energy must be sacrificed in order to obtain more with the cooling water.)

Closure

In illustrating the role played by available energy analysis in design optimization, one of two methods was demonstrated. The boiler problem used available energy balances to obtain all the pertinent consumptions and transports of available energy. This method reveals the relative importance of consumptions and losses with respect to the other transports of available energy into, out of, and within the system. It also provides a consistent basis for accurate costing of flow streams.

The error in energy analyses is that they attribute all the inefficiencies to losses, and then mis-calculate those. As was demonstrated for the coal-fired boiler, the first law efficiency (85%) was a poor approximation to the true efficience (33.8%). (Furthermore, perturbation studies show that the trends in first and second law efficiencies can move in opposite directions. For the coal-fired boiler problem, if the steam conditions were changed to $811^{\circ}K/6.87$ MPa ($1000^{\circ}F/1000$ psia) and if the first law efficiency were decreased to 83%, the result would be an increase to 34.3% in second law efficiency.) The major inefficiencies were due to heat transfer ($\eta_{II} = 47.3\%$) and combustion ($\eta_{II} = 73.8\%$), with the stack losses accounting for only 5.5% of the available energy input with the coal. In contrast, an energy analysis shows the combustion process to be 100% efficient and

the heat transfer process 98%, while the stack gases carry away
the majority of the lost energy. A designer relying on energy
analysis might try to lower the stack gas temperature in an attempt to recover some of the escaping sensible and latent heat
(and consequently waging war with sulfuric acid condensation).
Were he to convert to available energy accounting, it would be
clear that there is much greater potential for improvement by
increasing the steam temperature and pressure in the design or
by seeking   technologies with which to replace combustion,
rather than by trying to extract the last bit of available energy
out of the stack gases.

   Another important point that should be made is this: it is
misleading to imply that the value of a fuel lies in its heating
value. The true measure of a fuel's potential to cause useful
change for us is its content of available energy. This fact leads
to interesting "discrepancies." For instance, if an ideal power
plant were used in the boiler problem, so that the available energy in it were turned completely into electricity, that plant's
"thermal efficiency" (η_I) would be

$$\frac{\text{electrical energy output}}{\text{coal energy input}} = \frac{26390}{25607} \times 100 = 103\%!$$

This result stems directly from the fact that the coal's potential
energy content is 3% greater than its higher heating value.

   If in turn that electricity were used to drive a heat pump,
the maximum rate of heat deliverable at $305^{\circ}K$ ($90^{\circ}F$) with $T_0 =$
$278^{\circ}K$ ($40^{\circ}F$) would be $\dot{Q} = \dot{A}_{elec}/[1 - T_0/T]$, where T is the temperature at which the heat is delivered. In the limiting case of
perfect conversion the amount of heat energy delivered would exceed the amount of electrical energy consumed by a factor of
$[1/(1 - 278/305)] = 11.3!$ Even using second law efficiencies
typical of today's heat pumps and power plants, the heat delivered
by the heat pump would exceed the calorific input to the power
plant (as measured by the coal's heating value) by 35%! The point
here is not to promote heat pumps but to point out that it is
available energy, not energy, which measures a commodity's potential to effect changes, i.e., its value as a fuel.

   Available energy analysis can be applied as easily to a device (boiler) or process (heat transfer, combustion) as to an
overall system (power plant). Strategic use of this fact enabled
the consumptions within the coal-fired boiler to be pinpointed
and evaluated.

   As the impact of fuel cost on product cost continues to increase, so will the desirability of using available energy analysis in optimization of process designs. Efforts over recent years
have resulted in many simplified methods for evaluating transports
and destructions of available energy, which can be adapted for use
on computers or programmable calculators. It is hoped that, with
the removal of economic and theoretical barriers, second law

analysis will become a tool familiar to every designer of energy
conversion processes.

List of Symbols

$\dot{A}$ = available energy per unit time
a   = specific available energy
C   = weight fraction carbon in raw coal
c_p = heat capacity at constant pressure
$\dot{E}$ = energy per unit time
h   = specific enthalpy
H   = weight fraction Hydrogen in raw coal
HHV = higher heating value of raw coal
I   = current (commodity per unit time)
LHV = lower heating value of dried coal per unit moist coal
M   = moisture content (kg of H_2O/kg of coal)
$\dot{m}$ = mass flow rate
N   = weight fraction Nitrogen in raw coal
O   = weight fraction Oxygen in raw coal
p   = pressure
$\dot{Q}$ = thermal energy per unit time
R_0 = universal gas constant
s   = specific entropy
S   = weight fraction Sulfur in coal
T   = temperature
$\bar{v}$ = average specific volume
$\dot{W}$ = power; work rate
x   = mole fraction

Greek Symbols

η_I = first law efficiency
η_{II} = second law efficiency; true efficiency
μ = chemical potential
τ = torque
ω = angular velocity

Subscripts and Superscripts

α = angular momentum
δ = destruction
A   = air
C   = coal
f   = fluid
F   = feedwater
f_g = fluid/gas transition

```
G    = stack gases
G'   = dispersed stack gases
HT   = due to heat transfer
j    = jth component
L    = thermal energy loss
L'   = dissipated thermal energy loss
P    = products of combustion
RXN  = due to combustion
S    = steam
```

Literature Cited

1. Reistad, G; Gaggioli, R. "Available Energy Costing," this symposium.
2. Gaggioli, R.; Wepfer, W. "Available Energy Costing - A Co-generation Case Study," submitted to Chem Energy. (presented at the 85th National Meeting of AIChE, Philadelphia, PA, June 8, 1978).
3. Tribus, M.; Evans, R. UCLA Report No. 52-63, 1962.
4. Obert, E.; Gaggioli, R. "Thermodynamics," 2nd ed., New York: McGraw-Hill, 1963.
5. Gaggioli, R. "Principles of Thermodynamics," this symposium.
6. Rodriguez, L. "Calculation of Available Energy Quantities," this symposium.
7. Wepfer, W.; Gaggioli, R.; Obert, E. "Proper Evaluation of Available Energy for HVAC", ASHRAE Paper No. 2524 (to appear in ASHRAE Transactions, 1979.)
8. Hatsopoulous, G.; Keenan, J. "Principles of General Thermodynamics," New York: Wiley, 1965.
9. Obert, E. "Concepts in Thermodynamics," McGraw-Hill, 1960.
10. Wepfer, W.; Gaggioli, R. Reference Datums for Available Energy," this symposium.
11. Obert, E. "Thermodynamics," New York: McGraw-Hill, 1948.
12. Gaggioli, R.; Petit, P. Chemtech, 1977, 7, 496.
13. Gaggioli, R.; Yoon, J.; Patulski, S.; Latus, A.; Obert, E. "Pinpointing the Real Inefficiencies in Power Plants and Energy Systems," (presented at the American Power Conference, Chicago, IL, April, 1975).

RECEIVED October 19, 1979.

Calculation of Available-Energy Quantities

LUIS RODRÍGUEZ, S.J.

Department of Mechanical Engineering, Marquette University,
1515 W. Wisconsin Ave., Milwaukee, WI 53233

This paper deals with operational methods employed by the author in evaluating the available energy of various substances or mixtures thereof. Starting with availability equations valid for ideal-gas mixtures, some techniques employed in calculating the various contributions to the total available energy are explained. A correction for real-gas mixtures is then applied to the ideal values obtained for the mixture itself; such correction is based on Kay's mixture rule and generalized deviation charts..

The presence of liquid phases of some components of the mixture is then considered and a method to estimate the amounts of liquid phase of each component is discussed. The resulting liquid mixtures are treated as ideal solutions without consideration of any availability loss due to mixing. A formula for the availability of a moist air mixture is also discussed.

Some partial contributions to the total availability are then considered. Thus, a simple expression for the pressure availability of an incompressible fluid is developed. Formulae for the chemical availability of hydrocarbon fuels obtained by Szargut and Styrylska are then discussed and summarized in a separate table. Equations for the average value of the specific heat of various solid fuels between some fixed temperature and some other variable one are also given, as is a technique to estimate the lower heating value of a fuel of known atomic composition. Finally, a simplified approach used in approximating the thermal availability of tars is described.

Not all of the methods presented here have the same degree of reliability. Some, like the one used for tars, represent one possible attempt at bridging the gap created by lack of available experimental information and remain at the level of estimates. Others, like the formulae used for gas mixtures, are considered fully reliable for engineering analysis.

0–8412–0541–8/80/47–122–039$05.25/0
© 1980 American Chemical Society

Ideal Gas Mixtures

Gaggioli and Petit (1,2) discuss a formula for the availability transported per mole of a flowing mixture of ideal gases. The general expression

$$a(T,p) = h(T,p) - T_0 s(T,p) - \Sigma\, x_j \mu_{j,0} \qquad [1]$$

becomes for this particular case

$$a_{mix}(T,p) = \int_{T_0}^{T} \Sigma x_j c_{p_{j_{id}}} (1 - \frac{T_0}{T})dT + RT_0 \ln \frac{p}{p_0}$$

$$+ \Sigma x_j [h_{j_{id}}(T_0) - T_0 s_{j_{id}}(T_0, x_j p_0) - \mu_{j,0}] \qquad [2]$$

Without attaching strict physical meaning to the names to be now introduced, but using them simply as a communication help, the following terminology will be employed in this paper:

$$a_{j,t} \equiv \int_{T_0}^{T} c_{p_{j_{id}}} (1 - \frac{T_0}{T})dT \qquad [2a] \text{ thermal availability}$$

$$a_{j,p} \equiv RT_0 \ln \frac{p}{p_0} \qquad [2b] \text{ pressure availability}$$

$$a_{j,c} \equiv h_{j_{id}}(T_0) - T_0 s_{j_{id}}(T_0, p_0)$$

$$+ RT_0 \ln x_j - \mu_{j,0} \qquad [2c] \text{ chemical availability}$$

Since the term $RT_0 \ln \frac{p}{p_0}$ is a constant for any given state, and since $\Sigma x_j = 1$, it is possible to write

$$a_{j,p} = \Sigma x_j RT_0 \ln \frac{p}{p_0} = \Sigma x_j a_{j,p}$$

Then equation [2] may be written as

$$a_{mix}(T,p) = x_j(a_{j,t} + a_{j,p} + a_{j,c}) \qquad [3]$$

If an expression for the ideal specific heat at constant pressure $c_{p_{id}}(T)$ for each of the mixture's components is at hand, the calculation of $a_{j,t}$ can easily be programmed for computer use.

A short cut to evaluate this integration exists for gases included in available gas tables, since equation [2a] can readily be
converted into

$$a_{j,t} = h_j(T) - h_j(T_0) - T_0[\phi_j(T) - \phi_j(T_0)] \qquad [2a']$$

Both the ideal enthalpies and the ϕ-function, namely,

$$\phi(T) \equiv \int_{T_0'}^{T} \frac{c_p(T)}{T}\, dT \qquad [4]$$

are listed in most gas tables. Here the symbol T_0' is used for the
reference temperature employed in constructing a particular table;
it may or may not be the same as T_0 of the reference atmosphere.
     The contribution identified here as pressure availability
offers no difficulty. It is the evaluation of the chemical availability, $a_{j,c}$, that requires some explanation. The concept of
dead state is especially important.
     The availability of a material is the true measure of its
potential to cause change, as a consequence of not being completely stable relative to the ambient environment--as a consequence of
not being in a dead state relative to the reference atmosphere.
Although the concept of dead state as the zero-availability reference is discussed in other papers in this volume, along with
practical guidelines for selecting a reference environment, a
brief comment on the reference used in the present study will be
in order. The temperature and pressure of the reference are taken
to be, respectively, 298.15K and 1 atm. The reference atmosphere
is assumed to have the following various elements:
     1. For C, O and N their stable configurations are taken to
         be those of CO_2, O_2 and N_2, respectively, as they exist
         in air saturated with liquid water at (T_0,p_0).
     2. Hydrogen is assumed stable in the liquid phase of water
         saturated with air at (T_0,p_0). Thus the saturated liquid,
         not the saturated vapor at (T_0,p_0) is the reference state
         for water itself.
     3. The stable configuration of Ca is taken to be $CaCO_3$, cal-
         cite, at the reference environment conditions of T_0 and
         p_0.
     4. Finally, S is considered to exist in a stable configura-
         tion in the compound $CaSO_4 \cdot 2H_2O$, gypsum, at (T_0,p_0).
     Although the main criterion in selecting the stable configu-
ration for a given element is the thermodynamic stability of the
species selected, it is not the only one. The accessibility of
such species in the area for which a system is designed can and
will have an influence on the choice. Thus, for instance, the
reference environment chosen for calculations related to a system
to be installed in Antarctica will be different from that used for

an installation in the Sahara. It is, then, possible to select a given species as the stable configuration of a given element--for example, $CaCO_3$ for Ca--even if some other species, like $Ca(NO_3)_2$, can be proved to be thermodynamically more stable. A more detailed discussion of the "dead state" selection can be seen in Wepfer (4).

The values of $h_j(T_0,p_0)$ in equation [2c] may be evaluated directly with standard tabular values of (1) enthalpy of formation and (2) absolute entropy; alternatively, (3) Gibbs free energy of formation could be employed in lieu of the absolute entropy. What is crucial is this: Whichever pair is used to determine $h_j(T_0,p_0)$ and $s_j(T_0,p_0)$, the same pair should be employed to calculate μ_{j_0} (and, for that matter, the same pair should be used throughout the analysis of a process and/or system).

In computing the chemical potential $\mu_{j,0}$ of a component, a distinction has to be made between components that exist as stable species in the reference atmosphere and those that do not. For the first case,

$$\mu_{j,0}(T_0,x_{j,0}p_0) = g_j(T_0,x_{j,0}p_0) = h_j(T_0) - T_0 s_j(T_0,p_0)$$
$$+ RT_0 \ln x_{j,0} \qquad [5]$$

In this case, some of the terms present here will cancel out with other terms in equation [2c], and $a_{j,c} = RT_0 \ln (x_j/x_{j,0})$.

When the compound under consideration does not exist as a stable species in the reference environment, its chemical availability would be obtained by taking it to a completely stable configuration in the environment, employing only stable components from the environment to produce the appropriate stable component. Thus, for example, if the compound under consideration were CO, the appropriate reaction equation would be

$$CO + 1/2\ O_2 \rightarrow CO_2 \qquad or \qquad CO \rightarrow CO_2 - 1/2\ O_2$$

employing O_2 from the environment to convert CO to CO_2, which is a stable component of the environment. Then

$$\mu_{CO,0} = h_{CO,0} - T_0 s_{CO,0} = h_{CO_2}(T_0) - 1/2\ h_{O_2}(T_0)$$
$$- T_0 [s_{CO_2}(T_0,x_{CO_2,0}p_0) - 1/2\ s_{O_2}(T_0,x_{O_2,0}p_0)] \qquad [5a]$$

And

$$\mu_{CO,0} = h_{CO_2}(T_0) - 1/2\ h_{O_2}(T_0) - T_0[s_{CO_2}(T_0,p_0)$$
$$- 1/2\ s_{O_2}(T_0,p_0)] + RT_0 \ln \frac{x_{CO_2,0}}{(x_{O_2,0})^{1/2}} = -51.2793 \text{ kJ/mol} \qquad [5b]$$

As indicated earlier, the values of $h_j(T_0)$ and $s_j(T_0,p_0)$ can be read directly from ideal-gas tables; either absolute entropies or entropies of formation gotten, say, from Gibbs free energies and enthalpies of formation. Equation [5a] and [5b] could be written using $g_i(T_0,p_0)$ rather than $h_i(T_0,p_0) - T_0s_i(T_0,p_0)$, and Gibbs free energies of formation could then be used to evaluate the g_i, provided the Gibbs free energy of formation would also be used to evaluate $h_j(T_0,p_0) - T_0s_j(T_0,p_0)$ in equation [2c]. In such thermochemical property calculations, it is permissible to use independent bases for two but no more than two independent specific properties; for example, enthalpies of formation and absolute entropies, or enthalpies of formation and Gibbs free energies of formation. As another alternative, base enthalpies (to be defined presently) can be used in equations [2c], [5a] and [5b], and base entropies or base Gibbs free energies.

The resulting chemical availability values from equations [2c], and [5a] or [5b] for various compounds treated as ideal gases are given by Gaggioli and Petit in their appendix (1), relative to the aforementioned reference components at 298.15 K and 1 atm. These have been complemented with some newly calculated values and are presented in table I. Although the table was initially intended for components of gaseous mixtures, two solid substances --graphite and rhombic sulfur--have been included as additional information. Insofar as they are pure substances, the term $RT_0 \ln x_j$ has no meaning in their case.

Table I includes not only the chemical availabilities of the compound but also what is here referred to as the base enthalpy. The base enthalpy is akin to the enthalpy of formation; the latter is the enthalpy of a compound (at T_0,p_0) relative to the elements (at T_0,p_0) from which it would be formed. The former, the base enthalpy of a compound, is the enthalpy relative to the stable components of the environment--i.e., relative to the dead state. (Indeed, if values of availability are established relative to stable species of the environment, it is only logical to use the same criterion for the enthalpy values used in energy analyses.) Thus, if a compound exists as a stable component of the reference atmosphere, its base enthalpy is zero by definition. Otherwise, it is non-zero; to use the same example as employed before for $\mu_{j,0}$ of a species which is absent from the environment, the base enthalpy of CO is given by

$$h_{b,CO} = h_{CO} - h_{CO_2} - 1/2\ h_{O_2}$$

where the enthalpies of the environmental components CO_2 and O_2 are to be evaluated at ambient conditions. Enthalpies of formation can, of course, be employed to evaluate the terms on the right-hand side, since the reference enthalpies (of the elements) will cancel out.

Notice that the base enthalpy of CO is the same as the "heat of combustion" for CO. However, for example, H_2S has a base

Table I. Base Enthalpies and Chemical Availabilities of Some Components

COMPONENT	BASE ENTHALPY (kJ/mol)	CHEMICAL AVAILABILITY (kJ/mol)
Ammonia	382.585	$2.478907 \ln x^*_{NH_3} + 337.861$
Benzene	3,301.511	$2.478907 \ln x_{C_6H_6} + 3,253.338$
Carbon (graphite)	393.505	410.535
Carbon Dioxide	0.0	$2.478907 \ln x_{CO_2} + 20.108$
Carbon Monoxide	282.964	$2.478907 \ln x_{CO} + 275.224$
Carbon Oxysulfide	891.150	$2.478907 \ln x_{COS} + 848.013$
Ethane	1,564.080	$2.478907 \ln x_{C_2H_6} + 1,484.952$ (1,480.739#)
Hydrogen	285.851	$2.478907 \ln x_{H_2} + 235.153$
Hydrogen Sulfide	901.757	$2.748907 \ln x_{H_2S} + 803.374$
Methane	890.359	$2.478907 \ln x_{CH_4} + 830.212$
Nitrogen	0.0	$2.478907 \ln x_{N_2} + 0.693$
Oxygen	0.0	$2.478907 \ln x_{O_2} + 3.948$
Phenol	3,122.226	$2.478907 \ln x_{C_6H_5OH} + 3,090.784$ (3,156.576#)
Sulfur (rhombic)	636.052	608.967 (601.160#)
Sulfur Dioxide	339.155	$2.478907 \ln x_{SO_2} + 295.736$
Water	44.001	$2.478907 \ln x_{H_2O} + 8.595$

* x represents the component's molal fraction; for pure substances, of course, lnx = 0.

Obtained using tabulated values of g^o_f, instead of $(h_b - T_o s^o_{abs})$.

enthalpy

$$h_{b,H_2S} = h_{H_2S} + h_{CaCO_3} + 2h_{O_2} - h_{H_2O} - h_{CaSO_4 \cdot 2H_2O} - hCO_2$$

Corrections for Real-Gas Mixtures

Equation [2] and subsequent equations are for ideal-gas models. The usual methods of correcting for real-gas behavior may be employed to evaluate the available energy as given by equation [1].

The gaseous part of the reference environment described above behaves ideally, so that no corrections need to be made to the $\mu_{j,0}$. Therefore, only the h and s of the mixture as a whole need to be corrected; viz. equation [1]. A convenient means for approximating these corrections is with generalized deviation charts or equations. In order to make their use possible, a mixture rule to obtain pseudo critical temperature and pressure is needed, and Kay's rule offers the greatest simplicity in its application. Thus,

$$T'_c = \Sigma \, x_j T_c \tag{6}$$

$$p'_c = \Sigma \, x_j p_c \tag{7}$$

Using then the reduced properties $T_r \equiv T/T'_c$ and $p_r \equiv p/p'_c$, the following deviations can be read off a chart for the enthalpy and entropy of the mixture:

$$\Delta_h \equiv \frac{h_{mix_{id}}(T,p) - h_{mix}(T,p)}{T'_c} \tag{8}$$

$$\Delta_s \equiv s_{mix_{id}}(T,p) - s_{mix}(T,p) \tag{9}$$

As a result, the corrected properties of the real-gas mixture are

$$h = h_{id} - T'_c \Delta_h \tag{10}$$

$$s = s_{id} - \Delta_s \tag{11}$$

When these corrections are incorporated into equation [1], the resulting expression for the availability of a real-gas mixture follows

$$a = a_{id} + T_0 \Delta_s - T'_c \Delta_h \tag{12}$$

Although in principle it is possible to program these corrections using generalized virial equations of the state with T_r and p_r as variables, this author has found the direct use of charts less

cumbersome in correcting the ideal values obtained from programmed
calculations.

Some Special Gas Mixtures

Not infrequently, the thermodynamic state of a mixture is
such that the coexistence of liquid and vapor phases is suspected.
In this situation it is important to determine the molal composition of the vapor phase in order to apply to it the formulae just
discussed, treating the liquid phase or phases separately. As an
engineering approximation, the following procedure has been used
to determine which components could possibly be in two phases and
in what respective amounts.

A first step rules out a two-phase existence for all those
components whose critical temperatures are lower than that of the
mixture. When the latter is lower, the component could conceivably exist in two phases. In the cases encountered by the author,
the number of potential two-phase components was normally one or
two and exceptionally as high as four. These cases were analyzed
as follows:

First, the individual saturation pressures at the mixture's
temperature were obtained. If

$$p_{sat_i} > p_{mix},$$

it follows that the partial pressure p_i is smaller than the corresponding saturation pressure and that, therefore, that component is entirely vapor. If

$$p_{sat_i} < p_{mix},$$

the reasoning must be carried on further. Since the saturation
pressures encountered were considerably smaller than the respective critical pressures, the vapor phase was assumed to behave
ideally and, hence,

$$x_{sat_i} = p_{sat_i}/p \qquad\qquad [13]$$

It is known that the total molal mass of the gas mixture at saturation can be expressed as

$$N = N_v + \Sigma\, x_{sat_i} N \qquad\qquad [14]$$

where N_v represents the combined molal mass of all components
known to be in the vapor phase only and the subscript i refers to
any component potentially present in two phases. To simplify the
illustration, it will be assumed that only two components (i =
1,2) can exist in two phases. Then,

$$N - N_v = x_{sat_i} N + x_{sat_2} N \qquad [15]$$

$$N = \frac{N_v}{1 - (x_{sat_1} + x_{sat_2})} \qquad [16]$$

Thus, it follows that the values

$$N_{sat_1} = x_{sat_1} N \qquad\qquad 17$$

and

$$N_{sat_2} = x_{sat_2} N \qquad\qquad 18$$

can be estimated.

If N_{sat_i} exceeds the actual molal mass of component i, that component exists in the vapor phase only. Indeed, since $x_i N = N_i < x_{sat_i} N$, the smaller molal fraction $x_i < x_{sat_i}$ reflects a partial pressure $p_i = x_i p$ smaller than the corresponding saturation pressure $x_{sat_i} p$. If, on the other hand, $N_{sat_i} < N_i$, then N_{sat_i} represents the molal amount of vapor phase of that component. The balance, $N_i - N_{sat_i}$, exists in the liquid phase. If more than one component exists in the liquid phase, the resulting liquid mixture is then treated as an ideal solution.

Another particular case of interest in treating mixtures of gases is that of moist air encountered in most ventilating and air conditioning situations. By using a familiar expression for the molal fraction of water vapor x_w as a function of the humidity ration w, namely,

$$x_w = \frac{1.6078w}{1.0 + 1.6078w} \qquad [19]$$

Wepfer, et al., (5) have developed the following dimensionless formula for the availability of the moist air mixture

$$\frac{a}{c_{p_a} T_0} = (1.0 + 1.825w) \left(\frac{T}{T_0} - 1.0 - \ln \frac{T}{T_0} \right)$$

$$+ 0.2857 (1.0 + 1.6078w) \ln \frac{p}{p_0}$$

$$+ 0.2857 \ln\left[\left(\frac{1.0+1.6078w_0}{1.0+1.6078w}\right)^{(1.0+1.6078w)} \left(\frac{w}{w_0}\right)^{1.6078w} \right] \qquad [20]$$

where subscript a refers to dry air and subscript 0 refers to conditions in the reference environment.

Availability of Incompressible Fluids

For a pure substance, equation [1] becomes

$$a = h - T_0 s - \mu_0 \tag{21a}$$

which may be rewritten as

$$a = (h - h_0) - T_0(s - s_0) + (h_0 - T_0 s_0 - \mu_0) \tag{21b}$$

The last term in parentheses corresponds precisely to the substance's chemical availability, to use the terminology introduced earlier. The other two terms in parentheses can be evaluated by means of proper thermodynamic relations leading to the expression

$$a = \int_{T_0}^{T} c_p (1 - \frac{T_0}{T}) dT + \int_{p_0}^{p} [v - (T - T_0) \frac{dv(T,p)}{dT}] dp + a_c \tag{21c}$$

The evaluation of these integrals from the reference state (T_0, p_0) to a state (T,p) may be conveniently carried out along a two-stage path: first isothermally to (T_0,p) and then isobarically to (T,p). Along the isothermal path the first integral is zero and the second becomes

$$\int_{p_0}^{p} vdp.$$

On the isobaric path the second integral is zero, while the first yields precisely the thermal availability a_t. Thus,

$$a = a_t + \int_{p_0}^{p} vdp + a_c \tag{22a}$$

The new integral term corresponds to the pressure availability, a_p, of equation [3]. In the incompressible fluid model this term is readily evaluated as

$$a_p = v(p - p_0) \tag{23}$$

To the extent that a given liquid may be portrayed as incompressible, its availability can be obtained from the expression

$$a_{liq} = a_t + v(p - p_0) + a_c \tag{22b}$$

Chemical Availability of Hydrocarbon Fuels

Many of the fuels ordinarily encountered in industrial applications are not pure substances but a mixture of an often large number of components. Szargut and Styrylska (6) obtained in 1964 a series of equations applicable to hydrocarbon fuels which deserves some detailed consideration. They point out that, whenever the low heating value (LHV) and the absolute entropy of a fuel at (T_0, p_0) are known, the computation of its chemical availability can readily be obtained from the equation

$$a_c = LHV + T_0 s_0 + RT_0 (x_{O_2} \ln \frac{p_{O_2,0}}{p_0} - \Sigma x_k \ln \frac{p_{k,0}}{p_0}) \qquad [24]$$

where x stands for a molal fraction and the subscript k refers to any of the products of combustion; the symbols $p_{O_2 0}$ and $p_{k,0}$ refer to the partial pressure of oxygen or of a product of combustion in the reference environment.

Using equation [24], the chemical availability of many fuels with known values of LHV and s_0 was computed by Szargut and Styrylska and a correlation established in terms of their atomic ratios or corresponding mass ratios. Three categories of fuels were studied, namely, solid fuels containing no sulfur, sulfur-bearing solid fuels, and fluid fuels.

For solid fuels not containing sulfur and with a low level of oxidation, that is, with $\frac{O}{C} \leq 0.5$, they found that the correlation equation

$$\frac{a_c}{LHV} = 1.0438 + 0.0158 \frac{H}{C} + 0.0813 \frac{O}{C} + 0.0471 \frac{N}{C} \qquad [25a]$$

yielded an average deviation of $\pm$ 0.46%. The corresponding expression in terms of mass ratios, valid for $\frac{o}{c} \leq 0.666$, is

$$\frac{a_c}{LHV} = 1.0438 + 0.0013 \frac{h}{c} + 0.1083 \frac{o}{c} + 0.0549 \frac{n}{c} \equiv \beta_i \qquad [25b]$$

If the atomic ratio $\frac{O}{C}$ exceeds the value 0.5, the correlation formula for solid sulfur-free fuels is

$$\frac{a_c}{LHV} = \frac{1.0438 + 0.0158\frac{H}{C} - 0.3343\frac{O}{C}(1+0.0609\frac{H}{C}) + 0.0447\frac{N}{C}}{1 - 0.4043\frac{O}{C}} \qquad [26a]$$

and the resulting average deviation is given as $\pm$ 0.79% for $\frac{O}{C}$ up to 2.0. In terms of mass ratios, the equivalent expression, valid for $\frac{o}{c} \geq 0.666$, is

$$\frac{a_c}{LHV} = \frac{1.0438 + 0.0013\frac{h}{c} - 0.4453\frac{o}{c}(1+0.0051\frac{h}{c}) + 0.0521\frac{n}{c}}{1 - 0.5385\frac{o}{c}} \equiv \beta_2 [26b]$$

(Note: In transferring from atomic ratios to mass ratios, Szargut and Styrylska list incorrect numerical coefficients. Values given here represent the correct coefficients.)

The reason given by Szargut and Styrylska for treating separately sulfur-bearing solid fuels is that the combustion product
SO_2 does not exist as a stable species in the reference environment; moreover, it has an extremely small partial pressure in atmospheric air and is thus very difficult to measure accurately.
In order to handle this group of solid fuels, they ignore the
bond availability of sulfur in the fuel; that is, they treat sulfur as a free species. Next they calculate the difference between
sulfur's chemical availability and its lower heating value, given
by

$$(a_c - LHV)_s = \frac{(513,159 - 297.056)\ kJ/kmol}{32.064\ kg/kmol} = 6,740\ \frac{kJ}{kg} \qquad [27]$$

(Note: The authors being discussed seem to base some of their
numerical values on a set of tables different from reference (3).
The value of a_c given here is the one listed on table I, resulting
from reference (3) values, while the LHV of sulfur has been taken
from Chemical Engineer's Handbook (7).) From this they reason that
the chemical availability of solid fuels containing sulfur can be
obtained from the expressions

$$a_c = (LHV)\beta_1 + 6,740s\ kJ/kg, \quad if\ \frac{O}{C} \leq 0.666 \qquad [28]$$

$$a_c + (LHV)\beta_2 + 6,740s\ kJ/kg, \quad if\ \frac{O}{C} > 0.666 \qquad [29]$$

where in both cases s represents the mass fraction of sulfur in
the fuel. Commonly encountered sulfur-bearing solid fuels (such
as bituminous coal, lignite and cokes) have rather low oxidation
levels with $\frac{O}{C} \leq 0.666$. Understandably, no average deviations are
given for equations [28] and [29] since they are based on a theoretically reasoned approximation rather than on a direct correlation of precisely calculated values for sulfur-carrying solid
fuels.

In the study of fluid fuels Szargut and Styrylska did not
consider any of the organic substances containing nitrogen because the nitrogen mole fraction in fluid fuels is ordinarily
very small. Furthermore, availability losses associated with
mixing were also ignored since they are insignificant compared
with the chemical availability of a fluid mixture. Thus, for instance, in the case of a gas mixture of ten components in equal
proportions, the decrease in availability due to mixing was calculated to be 5,700 kJ/kmol, while its chemical availability had
a value of at least 800,000 kJ/kmol; in this illustration, then,
the availability lost in the mixing process amounted to no more
than 0.71% of the chemical availability.

For liquid fuels, the correlation formula

$$\frac{a_c}{LHV} = 1.0374 + 0.0159\ \frac{H}{C} + 0.0567\ \frac{O}{C} + 0.5985\ \frac{S}{C}\ (1-0.1737\ \frac{H}{C}) \qquad [30]$$

yielded an average deviation of $\pm$ 0.38%. The corresponding equation for gaseous fuels is

$$\frac{a_c}{LHV} = 1.0334 + 0.0183 \frac{H}{C} - 0.0694 \frac{1}{C} \qquad [31]$$

with a resulting average deviation of $\pm$ 0.15%.

All these correlation formulae were obtained from fuels for which both the lower heating value and the absolute entropy at (T_0, p_0) are known. Because the latter is not known for the generally used solid and fluid fuels such as bituminous coal, lignite, cokes, oil, tar, and so on, the equations are offered as plausible approximations for these based on analogy of behavior. Two very important observations should be made in this respect.

In the first place, the model used in these equations presupposes that the effect of intermolecular and intramolecular interactions upon the ratio a_c/LHV depends only on the atomic ratios. This is clearly an assumption, and the good correlations obtained by Szargut and Styrylska show that the assumption is good at least for the fuels they considered--all of them pure substances. There is no apparent reason to suspect that the correlation would not hold acceptably well for mixtures such as coal, petroleum, and so forth.

The second observation is explicitly made by the two authors in their paper. All correlations between chemical availability and lower heating value in terms of either atomic or mass ratios were based on values calculated for dry fuels with no moisture content. However, in applying those equations to fuels whose heating values do not appear on available tables, it is quite possible that the experimental determination of such value would be achieved by burning moist fuel in a calorimeter without first drying the substance. Care should then be exercised not to use the heating value of moist fuel in formulae derived for dry fuels. Nevertheless, the lower heating value of the dried fuel $(LHV)_d$ can be established in terms of that of the wet fuel $(LHV)_w$ by means of the relation

$$LHV_d = LHV_w + 2,442w \ kJ/kg \qquad [32]$$

where w represents the humidity ratio (kg water per kg of moist fuel) and the quantity 2,442 (kJ/kg water) corresponds to the enthalpy of vaporization of water. Whenever dealing with moist fuels, the proper heating value LHV_d given by equation [32] should be used in all correlation formulae. Of course, if the fuel is dry, then w = 0 and $LHV_d = LHV_w$.

With this in mind, Equations [25]-[26] and [28]-[31] have been summarized in table II. There the symbol $(LHV)_d$ has been used throughout, and it should be understood as representing either the lower heating value of dry fuel or that of dried fuel per unit of total, moist fuel from which it is derived. The same

Table II. Chemical Availability of Hydrocarbon Fuels (based on Szargut and Styrylska's work)

SOLIDS	FUELS	$\dfrac{a_c}{(LHV)_d} = 1.0438 + 0.0158\,\dfrac{H}{C} + 0.0813\,\dfrac{O}{C} + 0.0471\,\dfrac{N}{C}$	[25a]	if $\dfrac{O}{C} \le 0.5$
		$\dfrac{a_c}{(LHV)_d} = \dfrac{1.0438 + 0.0158\,\dfrac{H}{C} - 0.3343\,\dfrac{O}{C}\left(1 + 0.0609\,\dfrac{H}{C}\right) + 0.0447\,\dfrac{N}{C}}{1 - 0.4043\,\dfrac{O}{C}}$	[26a]	if $\dfrac{O}{C} > 0.5$
	CONTAINING ONLY C, O, H, N	$\dfrac{a_c}{(LHV)_d} = 1.0438 + 0.0013\,\dfrac{h}{c} + 0.1083\,\dfrac{o}{c} + 0.0549\,\dfrac{n}{c} \quad \left[\equiv \beta_1\right]$	[25b]	if $\dfrac{o}{c} \le 0.666$
		$\dfrac{a_c}{(LHV)_d} = \dfrac{1.0438 + 0.0013\,\dfrac{h}{c} - 0.4453\,\dfrac{o}{c}\left(1 + 0.0051\,\dfrac{h}{c}\right) + 0.0521\,\dfrac{n}{c}}{1 - 0.5385\,\dfrac{o}{c}} \quad \left[\equiv \beta_2\right]$	[26b]	if $\dfrac{o}{c} > 0.666$
	FUELS CONTAINING ALSO S	$a_c = (LHV)_d\,\beta_1 + 6{,}740\,s \quad \text{kJ/kg}$	[28]	if $\dfrac{o}{c} \le 0.666$
		$a_c = (LHV)_d\,\beta_2 + 6{,}740\,s \quad \text{kJ/kg}$	[29]	if $\dfrac{o}{c} > 0.666$
FLUIDS	LIQUIDS	$\dfrac{a_c}{(LHV)_d} = 1.0374 + 0.0159\,\dfrac{H}{C} + 0.0567\,\dfrac{O}{C} + 0.5985\,\dfrac{S}{C}\left(1 - 0.1737\,\dfrac{H}{C}\right)$	[30]	
	GASES	$\dfrac{a_c}{(LHV)_d} = 1.0334 + 0.0183\,\dfrac{H}{C} - 0.0694\,\dfrac{1}{C}$	[31]	

NOTATION:

$\dfrac{H}{C},\ \dfrac{O}{C},\ \dfrac{N}{C},\ \dfrac{S}{C}$ = atomic ratios

$\dfrac{h}{c},\ \dfrac{o}{c},\ \dfrac{n}{c}$ = mass ratios

a_c = chemical availability

$(LHV)_d$ = lower heating value } of dry fuel

s = sulfur mass fraction } either of dry fuel or of dried fuel per unit moist fuel from which it proceeds

understanding applies to the mass fraction of sulfur, s, and to the chemical availability, a_c, in all equations. If $(LHV)_d$ refers to dry fuel, the two above symbols will correspond to a unit of dry fuel; if $(LHV)_d$ corresponds to dried fuel, they, too, will refer to dried fuel per unit of moist fuel from which it proceeded.

Before concluding the discussion of Szargut and Styrylska's work, it is of interest to mention that they, too, use the concept introduced here earlier as the base enthalpy of a substance, although they refer to it as its devaluation enthalpy. However, the only two values they list corresponding to the devaluation enthalpies of S and SO_2 are in excess of those given here in table I by an amount equal to 88.07 kJ/kmol. The fact that exactly the same difference is observed in both cases leads the present author to suspect that a different set of values for some of the enthalpies of formation may have been used in their calculations.

Availability of Coals, Chars and Ashes

Of the three contributions to availability intorduced earlier, only the thermal and chemical availability are of any significance when dealing with solids, since their thermodynamic behavior is independent of pressure in the pressure ranges ordinarily encountered.

The thermal availability can easily be obtained from the expression

$$a_t = \int_{T_0}^{T} c_p (1 - \frac{T_0}{T}) \ dT \qquad [33]$$

if the function $c_p(T)$ is known. This information, however, is not always available for every variety of coal or char, but approximate formulae have been offered in the pertinent literature.

A. Lee (8) has developed a formula that correlates the mean specific heat of coal between 70 F and T in terms of its volatile matter content as follows

$$\bar{c}_p = 0.17 + 1.1 \times 10^{-4}T + (3.2 \times 10^{-3} + 3.05 \times 19^{-6}T)V_m \qquad [34]$$

where T is to be expressed in degrees Fahrenheit and V_m represents the volatile matter in weight percent of dry (not ash-free) coal. The mean specific heat $\bar{c}_p$ is called by Lee the "mean pyro-heat" and accounts for the rise in temperature itself as well as for the energy increase of the endothermic reaction of pyrolysis. The resulting value of $\bar{c}_p$ has units of Btu/lb F and can, of course, be taken out of the integrand in equation [33].

An alternative to the analytical approach offered by Lee is found in the tabulated values of mean specific heat between 0 and T degrees C, at different values of volatile matter content, published by the Institute of Gas Technology (9). In contrast to Lee's formula, the volatile matter content is here expressed as a

mass fraction of the dry, ash-free (not just dry) coal at 0 C.
The author has obtained nearly equal results using both approaches,
but has found the use of tables more straightforward.

Since the mean specific heat was used in cases where the reference environment was taken to be at 298.15 K, not at either 0 C
or 70 F, the following transformation of equation [33] was used

$$a_t = \bar{c}_{p_{0-T}} \int_{273.15\ K}^{T} (1 - \frac{T_0}{T})\ dT$$

$$- \bar{c}_{p_{0-25}} \int_{273.15\ K}^{298.15\ K} (1 - \frac{T_0}{T})\ dT \qquad [35]$$

The above formula holds also for ash, but its mean specific heat
between 0 and T degrees C is given by a different expression
found also in the Institute of Gas Technology report referred to
earlier. The formula

$$\bar{c}_{ash} = 0.18 + 7 \times 10^{-5}\ T\ (C)\ \frac{cal}{g\ K} \qquad [36]$$

holds in the temperature range 0 - 1,100 C.

The chemical availability of ash is assigned a value of zero
at the temperature and pressure of the reference environment.
For coals and chars, however, the pertinent formulae of table II
are used. The atomic (or mass) ratios can easily be obtained
from the coal's known composition. Its lower heating value can
be estimated with the following procedure used by the writer.

From the mass fractions of the different elements present in
the coal, a generic chemical formula can be established having the
form

$$C_a H_b O_d N_e S_f$$

where the subscripts represent atomic amounts per unit mass of dry
coal. The formula is then conveniently transformed to

$$CH_{b/a} O_{d/a} N_{e/a} S_{f/a}$$

per 1/a mass units of dry coal, and this fuel is considered as
consisting of a generic hydrocarbon $CH_{b/a} O_{d/a}$ plus free amounts
e/2a of N_2 and f/a of S.

These reactants are then burnt with the stoichiometric amount
of oxygen, and the enthalpy of reaction (high or low value) is
taken as the heating value of the coal. Admittedly, in evaluating
the enthalpy of reaction the (unknown) heat of formation of the
generic hydrocarbon itself is ignored and, thus, the results
represent an estimate, not a strict calculation of the heating
value.

Nevertheless, the procedure has been successfully tested

against more accurate techniques and has been found to be acceptably accurate. For example, in a 1973 report from the Bartles-
ville Energy Research Center (10), an experimental determination
of the lower heating value of a char identified as River King
Illinois No. 6 yielded a value of 9,940 Btu/lb of dry char, while
the above procedure applied to the same char yielded 9,930 Btu/lb.

The estimated value of the coal's (or char's) lower heating
value, together with the directly calculable atomic or mass
ratios of each element in the fuel, makes it possible to apply
the formulae of table II.

Availability of Tars

If the chemical composition of the tar is known, the procedure to compute its chemical availability is entirely parallel to
that used for coals and chars. After estimating the tar's lower
heating value and determining its atomic ratios, the corresponding
formula from table II is applied.

The pressure availability can be obtained by using the formula developed earlier in this paper for the incompressible fluid
model, namely

$$a_p = v\ (p - p_0) \qquad\qquad [23]$$

The value of v is easy to compute if the specific gravity of the
tar at hand is known. Because frequently this is not the case,
an average of representative specific gravity values for tars can
be used as an alternative since the contribution from either the
pressure availability, a_p, or the thermal availability, a_t, is
usually much smaller than the chemical availability, a_c.

The last remark serves also to obviate the difficulty of lack
of specific heat data for individual tars that prevents the exact
evaluation of the thermal availability from equation [33]. A
graphical procédure will now be described which was used by the
author as an approximation in a case where no information was
available on the specific heat of a tar present at some stages of
a coal gasification process. Specific heat data at seven temperatures for two different tar samples has been reported by Hyman and
Kay (11). A graph of the integrand of equation [33], $c_p(1 - T_0/T)$,
vs. temperature was prepared on large-sized graph paper, and both
sets of data plotted out clearly as straight lines, as shown in
Fig. 1. Both samples extrapolated to practically zero value of
the ordinate at $T = 25$ C, so that the graphical integration was
reduced to calculating the area of a triangle. Having graphically
evaluated equation [33] for both tar samples, the average value of
a_t was used as the thermal availability of the tar.

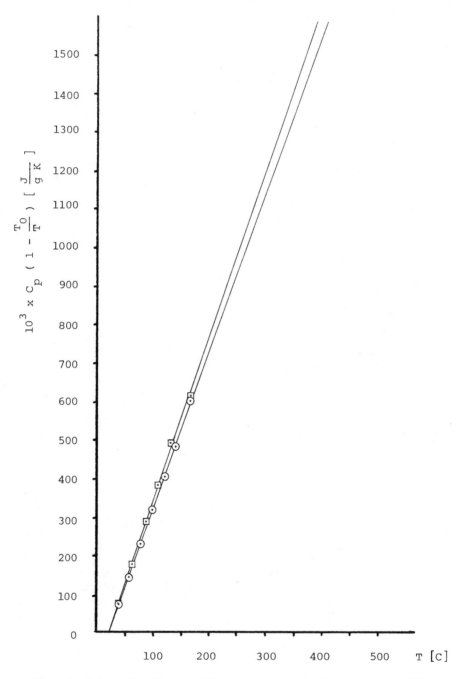

Figure 1. Integrand of Equation 33 vs. temperature for two tar samples: (□),
Sample 1; (⊙), Sample 2.

List of Symbols

a   - available energy (availability)

a,b,d,e,f   - atoms per unit mass

C   - Celsius; carbon atoms in fuel

c   - carbon mass in fuel; specific heat

F   - Fahrenheit

g   - gram; specific free energy

H   - hydrogen atoms in fuel

h   - hydrogen mass in fuel; specific enthalpy

J   - joule

K   - kelvin

k   - kilo-

LHV   - lower heating value

n   - nitrogen mass in fuel

O   - oxygen atoms in fuel

o   - oxygen mass in fuel

p   - pressure

R   - universal gas constant

S   - sulfur atoms in fuel

s   - sulfur mass ratio; specific entropy

T   - temperature

V   - volatile matter

v   - specific volume

w   - humidity ratio

x   - molal fraction

Greek Letters

β - defined by equations 8, 9

Δ - corrections defined by equations 25b, 26b

μ - chemical potential

ϕ - defined by equation 4

Subscripts

a   - of dry air

c   - at the critical point; chemical (availability) d   - of dry or dried fuel

f   - of formation

h   - of enthalpy

i   - of any component

id   - ideal gas

j   - of any component

k   - of a product of combustion

m   - matter

p   - at constant pressure; pressure (availability)

r   - pseudo reduced property

s   - of entropy; of sulfur

t   - thermal (availability)

v   - of vapor

w   - of moisture; of water vapor; of wet fuel

0(zero)   - of reference atmosphere

Superscripts

'   - property of mixture; table reference for ϕ-function

—   - mean

Literature Cited

1. Gaggioli, R. A.; Petit, P. J. "Use the Second Law First,"
    Chemtech, 1977, 7, 496-506.
2. Petit, P. J.; Gaggioli, R.A. "Second Law Procedures for
    Evaluating Processes," this volume.
3. Weast,R.C.,Ed. "Handbook of Chemistry and Physics," Chemical
    Rubber Company Press.
4. Wepfer, W. J. "Reference States for Available Energy
    Analyses," this volume.
5. Wepfer, W. J.; Gaggioli, R. A.; Obert, E. F. "Proper Eval-
    uation of Available Energy for HVAC," accepted by ASHRAE
    for presentation in Philadelphia, PA, January 28-February 1,
    1979, 37 pp.
6. Szargut, J.; Styrylska, T. "Angenäherte Bestimmung der
    Exergie von Brennstoffen," Brennstoff-Wärme-Kraft, 1964,
    16, 589-596.
7. Perry, R. H.; Chilton, C. H.; Kirkpatric, S. D., Eds.
    "Chemical Engineers' Handbook," McGraw-Hill Book Co.,
    1963, 4th ed.
8. Lee, A. "Heat Capacity of Coal," American Chemical Society,
    Chemical Preprints, 1968, 12, 19-31.

9.   Institute of Gas Technology,"Project 8964, Final Report,"
     Chicago, Illinois, 1976, 31 pp.
10. U. S. Department of the Interior, Bureau of Mines, Bartles-
     ville Energy Research Center, "Thermal Data on Gasifier
     Streams from Synthane Tests: Calculated Efficiency of Coal-
     to-Gas Conversion," Bartlesville, Oklahoma, November 1973,
     28 pp.
11. Hyman, D.; Kay, W. B. "Heat Capacity of Tars and Pitches,"
     Ind. Eng. Chem., 1949, 41, 1764-1768.

RECEIVED October 17, 1979.

Availability Analysis:
The Combined Energy and Entropy Balance

ELIAS P. GYFTOPOULOS

Massachusetts Institute of Technology, 77 Massachusetts Ave., Cambridge, MA 02139

THOMAS F. WIDMER

Thermo Electron Corporation, 101 First Ave., Waltham, MA 02154

The Measure of Efficiency

Availability analysis, i.e., the combined energy and entropy balance, has long been recognized as the most appropriate method for evaluating losses in, and efficiency of, processes. Despite this recognition, the use of availability for routine analyses of efficiency is far from commonplace. Apart from the power conversion field, in fact, many engineers still rely upon simple energy balances in their consideration of equipment and process design tradeoffs.

With rising energy prices and curtailment of supplies for certain high-quality fuels, there is a need to promote wider use of the concept of availability for all types of energy end-use assessments, both because it provides a better measure of the margin for improvement and because it discloses the real sources of inefficiency.

Currently, there is an enormous disparity between the most efficient and the least efficient processes and products in use today. A review of selected process and equipment efficiencies can provide a useful guide for establishing research and development priorities.

In general, availability analysis (as well as energy analysis) requires specifying the task to be performed, and evaluating availability changes of feedstocks and energy sources. Because of practical limitations, specifying a task is more often than not relative and not absolute and, therefore, availability analyses yield results that are relative to existing knowledge and technology.

For example, a common process encountered in industry is the heat treating of alloy steel parts to produce a locally hard surface (e.g., bearing or gear tooth wear surface). Though only a very small fraction of the material actually needs to be hardened, conventional technology has required that the entire part be heated to about 1650°F, then quenched at 350°F in oil to produce a hard martensite structure in the steel.

0–8412–0541–8/80/47–122–061$05.00/0
© 1980 American Chemical Society

One way of specifying the task is to say that the total mass of the part must be heated to 1650°F. Another way is to specify that only a small fraction (about 2% of the volume) of the material need be hardened. The availability change required by the first task is substantially different than that by the second. The results of the two availability analyses, however, are not comparable to each other because the tasks are different. They cannot be compared to each other, just as the task of making pig iron in a blast furnace cannot be compared to the task of making aluminum in an electrolytic cell.

In the example of steel hardening, the second specification of the task has, of course, little practical significance. However, the lower availability change required by the second task can provide useful guidance to R&D planners looking for innovative approaches to the problem of metal hardening. In fact, recent developments in high-power lasers and electron beam accelerators have led to the development of practical processes for localized heat treating (1). In one carburizing application, for example, electron beam heat treating has reduced total energy needed for a particular part from 3700 Btu to only 80 Btu. Thus, by redefining the task, the required availability was lowered well below the level previously thought to be ideal.

With regard to evaluations of availability changes (of either feedstocks or energy sources) we note that they require knowledge of the thermophysical properties and the initial and end states of the materials involved in the task. Such evaluations can be tedious. But in some widely used processes, a change in availability can be expressed as a product of a quality factor times the appropriate energy change, where the quality factor is a simple or tabulated function of some characteristic thermodynamic variable. Thus, the absolute thermodynamic efficiency η_α can be written in the form (2)

$$\eta_\alpha = \frac{C_2 \Delta E_2}{C_1 \Delta E_1} = \frac{\text{change in availability required by the task}}{\text{change in availability consumed by the task}}$$

where

ΔE_2 = energy required by the transformation induced by the task

C_2 = quality of the required energy

ΔE_1 = total primary energy (equivalent fuel energy) consumed in the task

C_1 = quality of the consumed energy

Quality is an important characteristic of energy because, as is well known, an amount of energy at high temperature is more useful and more valuable than an equal amount at low temperature.

Of course, any change in availability can be expressed as a product of a quality factor times an energy change, and values of efficiencies can be calculated for complete processes, such as the transformation of wood into a special type of paper, or of iron ore into a special steel. The results vary widely, even for slightly different types of the same product (e.g., hardened steel versus steel ingots). But values of efficiencies can also be determined for specific stages in a process (e.g., steam raising at specified conditions, or heat treating at a given temperature). For these stages, the results are numerically simple and are almost independent of the product (i.e., uniquely related to the equipment in question and its function). Examples of tasks that fall into the category just cited are:

- Heating of stock (such as steel parts) without chemical reactions or phase changes
- Raising of steam
- Generation of motive power and electricity
- Space conditioning (heating or air conditioning)

Each of these tasks requires an amount of energy, ΔE_2, of a certain quality, C_2. For a given level of output, the energy demand, ΔE_2, can be readily calculated by means of standard procedures. Its quality, C_2, on the other hand, can be evaluated as follows:

- If the task is generation of motive power or electricity,

$$C_2 = 1.0$$

- If the task is raising of steam at specific conditions

$$C_2 = 1 - 530 \frac{\Delta s}{\Delta h}$$

where Δs and Δh are the entropy and enthalpy changes from water at environmental conditions ($T_o = 70^o F$) to steam at the desired conditions.

- If the task is heating of stock at a particular temperature T ($^o F$)

$$C_2 = 1 - \frac{530}{T - 70} \ln \frac{T + 460}{530}$$

This expression is essentially the same as that for steam except that here Δs and Δh are readily expressible as functions of T by using the perfect gas relations.

Values of C_2 for certain steam conditions or for heating of stock are listed in Table I.

The energy required by each of the functions in question is satisfied by consuming fuel, electricity, or recovered waste heat. For each fuel, the energy, ΔE_1, is computed by using the heating value of the fuel. For electricity, the energy is the number of kWh consumed times 10,000 Btu per kWh because most electricity is generated from fuels at a rate such that one unit of electricity requires about 3 units of primary fuel energy. For waste heat,

the energy is found by considering changes in the energy content
of the material that carries the waste heat.

Table I.
Quality C_2 of Energy Demand
For Raising Steam and Heating Stock

Saturated Process Steam		Heating of Stock	
Pressure psia	Quality C_2	Temperature °F	Quality C_2
30	0.235	100	0.03
50	0.26	200	0.11
		300	0.17
100	0.295	400	0.22
		500	0.27
200	0.33	1000	0.42.
		1500	0.53
400	0.36	2000	0.58
		3000	0.66
600	0.38	4000	0.71

The quality factor, C_1, for most conventional fuels can·be
taken to be near 1.0 (i.e., availability is almost identical to the
heating value for petroleum, coal, and natural gas fuels). For
other fuels, such as low-Btu gas, waste materials, lignite, etc.,
the term $C_1\Delta E_1$ should be calculated on a case-by-case basis.
Where waste heat gases at constant pressure or steam are used as
input energy to a process, C_1 is determined in the same manner as
C_2 for output energy qualities (Table I).

The task-related efficiency can be evaluated either for a
single piece of equipment or for several pieces collectively. It
can also be applied to equipment with different types of materials
(outputs) being processed and several forms of energy being supplied. The energy-quality product for each output is evaluated as
discussed above and the results are additive. Moreover, the
energy-quality product for each of the energies supplied is computed as above and the results are additive. The overall efficiency is the ratio of the two sums.

It is noteworthy that the ratio $\Delta E_2/\Delta E_1$ instead of
$C_2\Delta E_2/C_1\Delta E_1$ is commonly used as a measure of efficiency for processes and devices. Though well defined, the ratio $\Delta E_2/\Delta E_1$
neither reveals the enormous opportunities for energy saving nor
addresses the real causes of inefficiency.

Illustrative Examples

Low-Temperature Processes. A residential water heater provides
a striking example of how a simplified availability analysis of
efficiency can be applied. Typically, a gas- or oil-fired water

heater will have an energy, or so-called "first-law" efficiency, of about 75% during operation; that is, the device will transfer about 3/4 of the fuel heating value into the enthalpy rise of the water. Due to standby losses, the overall in-service "efficiency" of the water heater is reduced to about 40%. Therefore, $\Delta E_2/\Delta E_1$ is 0.4.

Since domestic water heaters are required to deliver their output at only 140°F, the availability efficiency of such a unit is, of course, much lower than 40%. If the heater is fueled by a hydrocarbon, then $C_1 = 1.0$ and the quality, C_2, can be calculated by the expression for liquids or solids having near-constant specific heat; that is,

$$C_2 = 1 - (\frac{530}{140 - 70}) \ln \frac{140 + 460}{530} = 0.06$$

Thus, we find that the efficiency of the hydrocarbon-fired water heater is

$$\eta_\alpha = (\frac{C_2}{C_1})(\frac{\Delta E_2}{\Delta E_1}) = (\frac{0.06}{1.0})(0.4) = 2.4\%$$

If the water heater is electric, then

$$\frac{\Delta E_2}{\Delta E_1} = 0.85 = \frac{\text{enthalpy delivered to water}}{\text{electricity input}}$$

This ratio, which includes standby losses, is much higher than that for a hydrocarbon-fired unit because there is no heat leak through an exhaust flue. On the other hand, each unit of electricity input requires about 3 units of primary fuel and, therefore, with respect to the fuel consumed by the electric utility,

$$\frac{\Delta E_2}{\Delta E_1} = \frac{0.85}{3} = 0.28$$

Thus, we find that the efficiency of the electric water heater is

$$\eta_\alpha = (\frac{C_2}{C_1})(\frac{\Delta E_2}{\Delta E_1}) = (\frac{0.06}{1.0})(0.28) = 1.7\%$$

Equally small are the efficiencies of many industrial processes, particularly those involving low-temperature drying of materials such as food, textiles, or paper. This can be seen from an analysis of the efficiency of a dryer oven used to remove moisture from tobacco slurry. In this process, natural gas fuel is

burned to produce hot combustion gases. These gases are then
diluted with large quantities of excess air to provide drying gas
at a temperature of $330^{\circ}F$. The drying gas impinges directly onto
a moving conveyor carrying a thin sheet of tobacco-water slurry.
In a typical installation, 5700 lb/hr of moisture is removed at a
temperature of $130^{\circ}F$ (or 1080 Btu/lb) by burning 15,000 ft^3/hr of
natural gas (or 15.3 x 10^6 Btu/hr).

The "first law" efficiency of this process is

$$\frac{\Delta E_2}{\Delta E_1} = \frac{5700 \text{ lb/hr} \times 1080 \text{ Btu/lb}}{15000 \text{ ft}^3\text{/hr} \times 1020 \text{ Btu/ft}} = 0.4$$

On the other hand, considering the qualities of the energies used
and required at $130^{\circ}F$, we find

$$C_1 = 1.0$$

$$C_2 = 1 - 530 \left(\frac{\Delta s}{\Delta h}\right) = 1 - 530 \left(\frac{1.8367}{1080}\right) = 0.1$$

and

$$\eta_{\infty} = \left(\frac{C_2}{C_1}\right)\left(\frac{\Delta E_2}{\Delta E_1}\right) = \left(\frac{0.1}{1.0}\right)(0.4) = 4\%$$

The principal reason for this low efficiency is the availability
loss caused by using high-quality fuel to perform a task that requires only low-grade energy.

The dryer example provides an excellent illustration of the
opportunity that exists for both fuel substitution and fuel savings in most low-temperature industrial heating processes. It is
often claimed that gas fuel is absolutely necessary for many processes requiring a clean environment in ovens, kilns, crop dryers,
etc. This statement is true only for very high-temperature processes where heat exchanger problems might preclude the separation
of combustion products from the stock-heating media. But for low-
temperature applications, it is possible to use a separate combustion system, burning almost any type of fuel, and to heat clean
air (through a heat exchanger) for delivery to a process oven or
dryer. In addition, a topping engine can be installed for motive
power or electricity with the required heat exchanger attached to
the exhaust of the engine. In this way, the clean process environment is retained and efficiency is improved manyfold.

For the tobacco dryer system previously cited, it would be
possible to provide the 15.3 x 10^6 Btu/hr of heated air by re-
covering waste heat from the exhaust gases and water jacket of a
diesel engine generator set. Fuel effectiveness of the system is
increased because a substantial part of the fuel availability is

usefully employed to make byproduct electricity, and only low-
grade exhaust heat is employed for the actual drying operation.
    An energy balance of the system with the topping engine shows
the following distribution of outputs (relative to fuel input):

$$\Delta E_1 = \text{energy input}$$

$$\Delta E_2^{\ I} = 0.45\Delta E_1 \qquad = \text{energy to process air}$$

$$\Delta E_2^{\ II} = (0.4)(0.45)\Delta E_1 = \text{energy to evaporation process}$$

$$\Delta E_2^{\ III} = 0.35\Delta E_1 \qquad = \text{electricity}$$

$$\Delta E_2^{\ IV} = 0.20\Delta E_1 \qquad = \text{losses}$$

Thus, the availability efficiency is

$$\eta_\alpha = \frac{c_2^{\ II}\Delta E_2^{\ II} + c_2^{\ III}\Delta E_2^{\ III}}{c_1 \Delta E_1}$$

$$= \frac{(0.1)(0.4)(0.45)(\Delta E_1) + (1.0)(0.35)(\Delta E_1)}{(1.0)(\Delta E_1)} = 36.8\%$$

Total fuel consumption of the combined system will be 34 x 10^6
Btu/hr, an increase of 18.7 x 10^6 Btu/hr over conventional practice. Electricity output, however, amounts to 3500 kW, and this
represents an incremental fuel rate of 18.7 x 10^6/3500 = 5340 Btu/
kWh. Thus, there is a saving of at least 4000 Btu/kW relative to
electricity produced by central station powerplants.
    Obviously, the topping engine approach will cost far more
than a simple, once-through system using only a gas burner with
air dilution ports. Also, means must be found to use the byproduct electricity in other operations at the same site, or to deliver the power to a utility grid. The critical consideration
from a national viewpoint, however, is the fact that this conservation measure will yield much more energy per dollar of capital
invested than will comparable investments in new energy supply.
    In another type of dryer, steam is used to transfer heat into
large steel rolls over which paper is passed at high speed to remove moisture added in the forming process (3). Typically, the
paper web temperature is held to about 170°F. Steam is introduced
to the rolls at about 50 psig, with about 2.2 pounds of steam
needed to remove each pound of moisture. Process boilers are commonly fired by either residual oil or coal, augmented by waste
material such as wood chips and spent pulp liquors.
    In a large machine, moisture is removed at a rate of about
30,000 lb/hr, necessitating a steam flow of 66,000 lb/hr. Boiler
efficiency (first law) is usually about 0.88; therefore, the fuel
energy requirement is

$$\Delta E_1 = 66{,}000 \text{ lb/hr} \times 1136 \text{ Btu/lb} \times \frac{1}{0.88}$$

$$= 85 \times 10^6 \text{ Btu/hr}$$

For the boiler, C_1 is approximately 1.0 and C_2 (from Table I) is 0.26 and therefore,

$$\eta_B = \frac{C_2 \Delta E_2}{C_1 \Delta E_1} = (\frac{0.26}{1.0})(0.88) = 23\%$$

The dryer output (ΔE_3) is represented by the moisture removed.

$$\Delta E_3 = 30{,}000 \text{ lb/hr} \times 1096 \text{ Btu/lb} = 32.9 \times 10^6 \text{ Btu/hr}$$

the input steam quality factor is $C_2 = 0.26$, and the output moisture quality C_3 is given by

$$C_3 = 1 - 530 \; (\frac{1.7548}{1096}) = 0.15$$

Hence, dryer efficiency is

$$\eta_D = \frac{C_3 \Delta E_3}{C_2 \Delta E_2} = \frac{(0.15)(32.9 \times 10^6)}{(0.26)(85 \times 10^6 \times 0.88)} = 25.5\%$$

Overall system efficiency is the product of the boiler and dryer efficiencies:

$$\eta_\alpha = \eta_B \times \eta_D = 0.23 \times 0.255 = 5.8\%$$

Once again, the advantages of a topping cycle are obvious. In practice, many paper mills use high-pressure boilers with back-pressure steam turbines to produce about 50 kW of electricity for every 1000 lb/hr of process steam. Much greater electrical output and higher overall efficiency could be realized if a portion of the steam were supplied by using diesel engines with recovery boilers. Large two-stroke diesel engines can burn residual fuel. If used in the example shown here such engines would be able to provide as much as 26,000 kW of byproduct electricity. Overall efficiency of the process would rise from 5.8% for the simple system to 16.1% for the back-pressure turbine topping cycle, and to 35.3% for the diesel topping cycle.

Vehicle Performance. Efficiency of a vehicle can also be calculated directly from input-output considerations. As we might expect, its value depends on the definition of the task to be performed. If, for example, the task is defined as moving X

passengers from point A to point B without change in elevation,
then the overall efficiency of a typical automobile, based on
either energy or availability changes, would be zero. Though correct, such a result is of no practical use.

Another way of defining the task might be to establish a certain mass and frontal area of vehicle needed to obtain a particular level of comfort and safety, and to set specifications for the
rate of speed and acceleration desired. Then the availability required for moving X passengers from point A to point B can be calculated from the rolling resistance and drag forces and compared
to the availability consumed.

A still different approach would be to use the actual figures
for drag and rolling resistance of a specific vehicle (e.g., 1978
Chevrolet Sedan weighing 4200 pounds). The actual horsepower delivered to the road can then be integrated over the time of a
specific driving cycle (e.g., EPA Metro-Highway program) or running at a constant speed. Dividing this actual work done by the
fuel availability consumed over the test period gives an average
efficiency for the vehicle.

This might be called a "machine efficiency," since it ignores
any possibilities for improvement which might result from modify-
ing the task. For example, reducing vehicle weight by using
lighter materials or lowering aerodynamic drag by streamlining
would not change the calculated efficiency. The figure would be
affected, however, by any propulsion system improvements such as
reduced engine friction, lower accessory losses, or better match-
ing between engine and power train.

Applying the "machine-efficiency" definition to a typical
automobile operating at steady 50-mph speed on level road, the
only data required are the road load fuel consumption and total
drag figures. For a particular Ford Galaxie weighing 4576 lb, the
measured figures were 18.5 mpg at 50 mph with total drag (windage
and rolling) of 157.5 lb force. On the basis of these data we
find:

ΔE_2 = rate of output work

$$= \text{Drag x Velocity} = \frac{157.5 \text{ lb x } 50 \text{ mph x } (\frac{5280}{3600})}{550 \text{ ft \#/Sec hp}}$$

$$= 21 \text{ hp} = 53{,}500 \text{ Btu/hr}$$

ΔE_1 = rate of input fuel availability

$$= \frac{50 \text{ mph}}{18.5 \text{ mpg}} \text{ x } 6.1 \text{ lb/gal x } 18{,}500 \text{ Btu/lb}$$

$$= 305{,}000 \text{ Btu/hr}$$

C_2 = 1.0 because the output is mechanical work

C_1 = 1.0 because the availability of gasoline is almost equal to its heating value

and

$$\eta_\propto = (\frac{1.0}{1.0})(\frac{53,000}{305,000}) = 17.5\%$$

This efficiency is relatively high because 50-mph road load represents a near optimum condition for the engine and transmission. In-service efficiencies will generally be lower because of both off-optimum running and losses incurred from acceleration and braking. Regenerative braking (instead of friction) is one possible method for reducing the work loss in a vehicle that must operate with frequent stops and starts.

Heavy diesel trucks have considerably higher efficiency than automobiles, due to the greater efficiency of diesel engines relative to spark ignition engines, and because of the lack of significant light-load operation during normal running.

Test data for a 72,000-lb (gross combination weight) tractor-trailer illustrate this point. The tractor is powered by a 676 in.^{3} turbo-charged and intercooled Mack diesel. When driven over a standard test route (NAPCA Control Route), the engine output as a percentage of time is that given in Table II.

Table II.
Duty Cycle for Mack Diesel Tractor-Trailer
Over NAPCA Control Route Driving Cycle

Horsepower	Percent time at given horsepower
Below 165	15.1
165 to 225	1.6
225 to 270	9.5
270 to 300	73.0

Average work delivered to the road is 261 hp over the entire cycle, with a fuel consumption rate of 104 lb/hr (an average of about 3.5 mpg). Thus, vehicle efficiency is given by

$$\eta_\propto = \frac{C_2 \Delta E_2}{C_1 \Delta E_1} = \frac{(1.0)(261)(2546)}{(1.0)(104)(18,500)} = 34.5\%$$

This represents a machine efficiency roughly equal to that of central station electric power production and distribution. Even higher efficiency can be obtained with a new experimental compound engine system. Known as an organic Rankine bottoming cycle, this powerplant derives about 38 hp (at full load) from the waste heat of the truck engine. Organic fluid is vaporized in an

exhaust gas boiler, expanded through a small turbine, and condensed in a heat exchanger cooled by the truck radiator. Turbine
power is geared directly into the transmission and thus augments
the diesel output. Laboratory tests of the complete powerplant
have demonstrated an improvement of 13% in fuel economy over the
simulated driving cycle and, therefore,

$$\eta_\alpha' = (0.345)(1.13) = 39\%$$

Applying this improvement to just the long-haul segment of the
heavy truck fleet would yield savings of over 100,000 barrels of
distillate fuel per day.

   Chemical Processes. For complex chemical or metallurgical
processes, the evaluation of availability flows requires more
elaborate calculations.
   The production of ammonia from natural gas (methane) provides
an illustrative example. In a paper by L. Rieker (4), the common
process for ammonia production is represented by the schematic of
Figure 1. Material flows between process units is defined by the
vertical bars, the width of the bars indicating relative magnitude
of the availability.
   Table III summarizes the availability values of each stream,
prorated on the basis of each ton of ammonia output. The loss at
each stage of the process is given in Table IV. Overall efficiency of the complete process is given by

$$\eta_\alpha = \frac{\text{Availability Content of Ammonia Output}}{\text{Availability Content of Methane Input}}$$

$$= \frac{17.5 \times 10^6 \text{ Btu/ton}}{31.4 \times 10^6 \text{ Btu/ton}} = 0.56$$

Thus, well over half of all the availability contained in the
methane feedstock is contained in the ammonia produced by this
process. An alternative process for making ammonia from water and
air, using electricity produced from coal, has an overall efficiency of only about 17%.
   Another example of a process involving chemical transformation is the blast furnace used in converting iron ore (Fe_2O_3) into
molten iron (5). In fact, the blast furnace is not simply a furnace but is actually a highly efficient counterflow thermochemical reactor. The enthalpy and availability figures for a
typical blast furnace (Table V) show that 75% of the availability
contained in feed materials is preserved in the output iron and by-
product fuel gases (principally CO).

Table III.
Availability Content of Process Stream
in Ammonia Plant

Process stream (numbers refer to Figure 1)	Availability (Btu per ton of NH_3)
(1) methane input	31.4×10^6
(2) intermediate mixtures	24.7
(3) intermediate mixtures	23.6
(4) intermediate mixtures	21.7
(5) H_2, N_2	21.4
(6) CO_2	0.16
(7) H_2, N_2	21.3
(8) purge gas	2.5
(9) NH_3 output	17.5
(10) Steam	4.46
(11) Steam	1.42
(12) Steam	0.1

Table IV.
Distribution of Losses
in Ammonia Process

Process stage	Loss of availability (Btu per ton of NH_3)	Stage Efficiency
Primary Reformer	4.79×10^6	0.85
Secondary Reformer	1.10	0.96
Shift Conversion	0.53	0.97
CO_2 Removal	0.14	0.99
Methanation	0.10	0.99
Compression Synthesis	7.08	0.74

Table V.
Enthalphy and Availability Balance
for Blast Furnace

(All Units 10^6 Btu/ton Pig Iron)

	Enthalpy	Available Useful Work
Process Fuels Consumed		
Inplant Coke	14.08	14.11
Purchased Fuels Consumed		
Merchant Coke	1.01	1.01
Injectants		
Natural Gas	0.33	0.30
Fuel Oil	0.58	0.56
Byproduct Fuels Consumed		
Injectants		
Coke Oven Gas	0.06	0.05
Tar Pitch	0.15	0.15
Blast Stove		
Blast Furnace Gas	1.61	1.50
Utilities		
Electricity	0.15	0.05
Steam	1.46	0.65
Oxygen	0.01	–
Total Fuels and Utilities	19.44	18.38
Byproduct Fuels Produced		
Blast Furnace Gas	6.57	5.9
Raw Material Output	8.25	7.85
Lost in Process	4.62	4.63
Process Effectiveness		75%

Appendix

The Combined Energy and Entropy Balance. Several approaches exist for establishing that availability change and not change in any other property represents the optimum (minimum or maximum) work requirement of a process. One of these approaches is based on a combination of the energy and entropy balances of the process.

The laws of thermodynamics imply the existence of two properties: energy and entropy. These properties are such that: (a) the energy of all systems involved in a process is conserved; i.e., the energy in any process must be balanced; and (b) the entropy of all systems involved in a process either increases or remains invariant; i.e., the entropy in any process must be balanced by considering a nonnegative amount of entropy due to irreversibility. The energy and entropy balances are essential to any thermodynamic analysis.

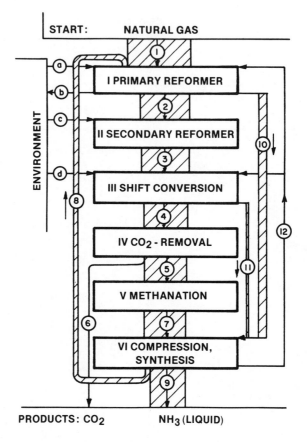

Figure 1.   Process flowsheet for conversion of methane to ammonia

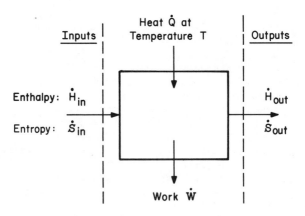

Figure 2.   Energy and entropy flows in a steady-state bulk flow process

The forms of the two balances depend on the particular circumstances of the process. As an illustration, we will consider a bulk flow process in steady state (Figure 2) in a chamber with a fixed volume. The energy and entropy rate balances are:

Energy rate balance

$$\dot{H}_{in} - \dot{H}_{out} + \dot{Q} - \dot{W} = 0 \tag{1}$$

Entropy rate balance

$$\dot{S}_{in} - \dot{S}_{out} + \frac{\dot{Q}}{T} + \Delta\dot{S}_{irr} = 0 \tag{2}$$

where $\Delta\dot{S}_{irr}$ denotes the entropy rate due to irreversibility.

Multiplying Eq.(2) by T_o, the temperature of the environment, and subtracting the result from Eq.(1) we find

$$\dot{W} - \frac{T - T_o}{T}\dot{Q} = \left[(\dot{H}_{in} - T_o\dot{S}_{in}) - (\dot{H}_{out} - T_o\dot{S}_{out})\right]$$

$$- T_o\Delta\dot{S}_{irr} \tag{3}$$

Clearly, $\left[(T - T_o)/T\right]\dot{Q}$ is the optimum work rate obtainable from a heat source at temperature T with respect to the environment at temperature T_o. The overall process will be optimum when the irreversibility is zero ($\Delta\dot{S}_{irr} = 0$) and, therefore, the optimum work rate is defined by the change in $\dot{H} - T_o\dot{S}$; i.e., the change in the availability rate of the bulk flow process.

Literature Cited

1. Obruzut, J.J., "Heat Treaters Gear Up for the New Demands," Iron Age Magazine, July 10, 1978.
2. Hatsopoulos, G.N., Gyftopoulos, E.P., Sant, R.W., and Widmer, T.F., "Capital Investment to Save Energy," Harvard Business Review, Vol. 56, No. 2, March-April 1978.
3. Villalobos, J.A., "The Effective Use of Energy in the Paper Drying Process," International Symposium on Drying, McGill University, August 3, 1978.
4. Riekert, L, "The Efficiency of Energy Utilization in Chemical Processes," Chemical Engineering Science, Vol. 29, 1974.
5. Gyftopoulos, E.P., Lazaridis, L.J., and Widmer, T.F., "Potential Fuel Effectiveness in Industry," Report to the Energy Policy Project of the Ford Foundation, Ballinger Publishing Co., 1974.

RECEIVED November 1, 1979.

Reference Datums for Available Energy

WILLIAM J. WEPFER

Professional Engineering Consultants, Milwaukee, WI 53211

RICHARD A. GAGGIOLI

Department of Mechanical Engineering, Marquette University,
1515 W. Wisconsin Ave., Milwaukee, WI 53233

The purpose of this article is to provide appropriate criteria for the practical selection of reference datums for the calculation of available energy. The selection of a reference datum generally depends on the commodity whose available energy is being evaluated, upon the particular process (or device) being analyzed, upon the complex of processes (devices) with which the particular process interacts, and upon the ambient environment of the complex.

Proper choice of a reference datum (a dead state) is important to efficiency analysis and costing. It needs to be recognized that the reference datum for available energy is an altogether different concept than the reference base for property tabulations.

The selection of a reference base for property tabulations is arbitrary. Whatever base is selected, the base values of the different extensive properties will cancel out of the thermodynamic property calculations--assuming, of course, that the calculations are carried out correctly. (That is, when employing thermochemical property tabulations to make property calculations, it is necessary to assure that the base values cancel out; this must be done for all extensive properties--enthalpy, entropy,..., availability, etc.)

Some contend that the chemical reference datum for available energy can also be selected arbitrarily, just like a base for thermochemical tables (while admitting that the thermal reference datum--the "dead state temperature"--is not arbitrary). The contention is erroneous; changing the various values of the available energy of a specific material by a constant amount (as a consequence of changing the reference datum) leads to misconceptions, to misevaluations, and to misallocations--in the determination of inefficiencies and costs. Absolute values of available energy can and should be evaluated.

Before proceeding to the criteria for practical selection of the dead state reference datum for analyzing a particular process, some background fundamentals will be presented.

0–8412–0541–8/80/47–122–077$05.00/0
© 1980 American Chemical Society

Theoretical Preliminaries

System Available Energy. The available energy as commonly defined (1) and symbolized by A is a special case of system available energy, B. (As will be seen, A is called the subsystem available energy in this paper.)

Given a system at an arbitrary state, with energy $E(t)$, entropy $S(t)$, volume $V(t)$, etc., the system available energy $B(t)$ is defined as the maximum amount of energy that could be delivered from the system by processes with no net transport of any other extensive property. Then

$$B(t) = E(t) - E_f(t) \qquad (1)$$

where $E_f(t)$ is the energy at the state f which, among all the states that could be reached with no net transport of any extensive property besides energy, has the minimum energy.

The energy, $E(t) - E_f(t)$, could be obtained via volume transports, via entropy transports or via any extensive property transports or combination of extensive property transports. This point is illustrated by several examples in (2).

Available Energy Destruction and Entropy Production. The Second Law of Thermodynamics can be stated to decree that the system available energy of an isolated system decreases in all real processes. Since $E(t)$ is constant for an isolated system it follows that $\dot{E}_f(t) \geq 0$.

Consider the system illustrated by Fig. 1, and described in the figure caption. The available energy of the whole system is represented by area $mf_1a'm$ on Fig. 2. If the system were isolated, then for any state 2 that could be reached spontaneously from state 1 of Fig. 1, the dead state would have

$$V_{I_{f_2}} = V_{II_{f_2}} = V_{f_2}$$

$$S_{I_{f_2}} = S_{II_{f_2}} = S_{f_1} + S_\pi/2$$

From the figure it can be seen that the dead state energy would be given by

$$E_{f_2} = E_{f_1} + \int T_f \dot{S}_\pi \, dt$$

$$= E_{f_1} + \int T_f dS_\pi$$

where S_π is the amount of entropy produced.

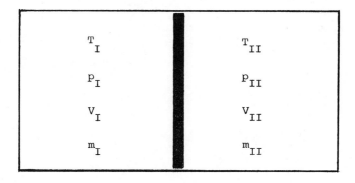

Figure 1.   *A system having* $T_{II} > T_I$, $S_{II} > S_I$, $p_{II} = p_I$ *and* $m_{II} = m_I$

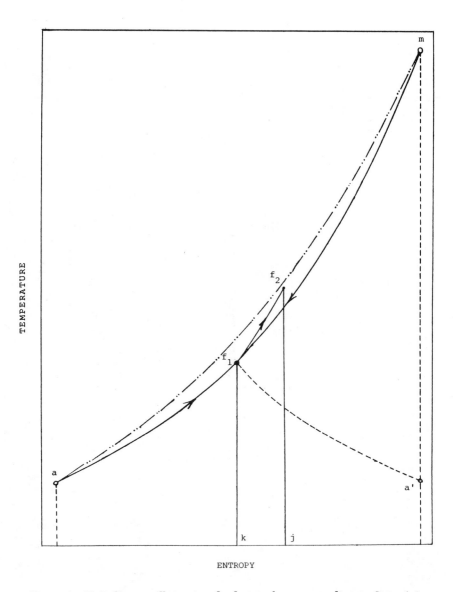

Figure 2. T–S *diagram illustrating dead state* f$_1$ *corresponding to State 1 in Figure 1. State* f$_2$ *is the dead state for some State 2 that could be reached spontaneously from State 1.*

Where the integral is twice the area $f_1f_2jkf_1$ shown on Fig. 2.
(Since $E_{f_2} \geq E_{f_1}$, the integral is equal to or less than area
$mf_1a'm$. The state f_2 which yields the maximum integral, equal
to area $mf_1a'm$ is the equilibrium state that the system would
reach from state f_1, were isolation maintained.)   Since $B = E - E_f$, it follows that

$$\dot{B}_{isolated}(t) = - \dot{E}_f(t) = -T_f(t)\dot{S}_\pi(t) \qquad (2)$$

<u>Subsystem Available Energies</u>.   The purpose of this section
is to show that subsystem available energies, A, can be defined
such that the system available energy, B, of any system is equal
to the sum of the subsystem available energies. That is, for
any breakdown of the system into distinct subsystems

$$B = \Sigma A_i \qquad (3)$$

where A_i is an <u>extensive property</u> of subsystem i ($B \neq \Sigma B_i$; i.e.,
B is not an extensive property.) Furthermore, it will be shown
that for any subsystem

$$A = E + p_f V - T_f S - \Sigma \mu_{if} N_i \qquad (4)$$

where p_f, T_f, μ_{if},..., are the pressure, temperature, and
chemical potentials ...of the subsystem at the dead state of the
composite (whole) system.
    Before proceeding to the above-mentioned developments, it
needs to be noted that the system available energy B_{IUII} of the
composite IUII of two systems I and II equals the sum of their
individual system available energies, $B_I + B_{II}$, plus the system
available energy $B_{I_fUII_f}$ of the composite when I is in its dead
state and II is in its dead state:

$$B_{IUII} = B_I + B_{II} + B_{I_fUII_f} \qquad (5)$$

    Consider the composite system shown in Fig. 3 where sub-
systems A and B can exchange entropy, volume, and components
i = 1,....   The available energy of A is the same as that of
AUC, where C is an infinite environment at $T_f, p_f, \mu_{if}...$, the
dead state temperature, pressure and chemical potentials of AUB.
Similarly for B and BUD. The available energy of the composite,
B_{AUB}, is the same as the available energy of the composite
R_1UR_2:

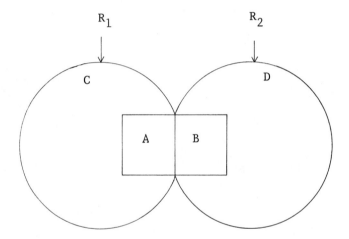

Figure 3. A composite system consisting of Subsystems A and B, at a state of the composite which has dead-state properties equal to T_f, p_f, and μ_{if}. . . . Subsystem B is surrounded by an infinite environment D having the same T_f, p_f, and μ_{if}. . . . Similarly for C, which surrounds A. Note that $R_1 = A \cup C$ and $R_2 = B \cup D$.

$$B_{AUB} = B_{R_1 U R_2}$$

$$= B_{R_1} + B_{R_2} + B_{R_{1_f} U R_{2_f}} \tag{6}$$

However, because both R_1 and R_2 are infinite in extent and have the same dead-state intensive properties $T_f, p_f, \mu_{if} \ldots$, it follows that $B_{R_{1_f} U R_{2_f}}$ is equal to zero. Furthermore, the available energy of R_1 is equal to the maximum energy that can be extracted from the composite AUC

$$B_{R_1} = \{E - E_f\} + \{E - E_f\}_C$$

Since the maximum energy is extracted with processes having $S_\pi = 0$, the term $\{E - E_f\}_C$ can be rewritten by using a form of the Gibbs Equation (3-7), $dE = T \, dS - p dV + \Sigma \mu_i dN_i$:

$$\{E - E_f\}_C = p_f \{V_f - V\}_C - T_f \{S_f - S\}_C + \Sigma \mu_{if} \{N_i - N_{if}\}_C$$

where N_i represents the components (2,4) of C. Thus

$$B_{R_1} = \{E - E_f\}_A + p_f \{V_f - V\}_C - T_f \{S_f - S\}_C + \Sigma \mu_{if} \{N_i - N_f\}_C \tag{7}$$

However, note that a volume balance on system R_1 (subsystems A and C) gives

$$\{V_f - V\}_C = -\{V_f - V\}_A$$

Similar balances can be written for entropy and components so that Eq. 7 can be written in the following form

$$B_{R_1} = \{E - E_f\}_A - p_f \{V_f - V\}_A + T_f \{S_f - S\}_A - \Sigma \mu_{if} \{N_i - N_{if}\}_A \tag{8}$$

Likewise

$$B_{R_2} = \{E - E_f\}_B - p_f \{V_f - V\}_B + T_f \{S_f - S\}_B - \Sigma \mu_{if} \{N_i - N_{if}\}_B \tag{9}$$

Substitution of Eqs. (8) and (9) into Eq. (6) yields

$$B_{AUB} = E_A + p_f V_A - T_f S_A - \Sigma \mu_{if} N_{iA}$$

$$- \{E_{A_f} + p_f V_{A_f} - T_f S_{A_f} - \Sigma \mu_{if} N_{iA_f}\} +$$

$$+ E_B + p_f V_B - T_f S_B - \Sigma \mu_{if} N_{iB}$$

$$-\{E_{B_f} + p_f V_{B_f} - T_f S_{B_f} - \Sigma \mu_{if} N_{B_f}\} \qquad (10)$$

From classical thermodynamics (4,6) it is known that a form of the Gibbs equation can be integrated to give $E = TS - pV + \Sigma \mu_i N_i$. Thus, the terms in the brackets in Eq. 10 are identically equal to zero. That is

$$B_{AUB} = E_A + p_f V_A - T_f S_A - \Sigma \mu_{if} N_{iA}$$

$$+ E_B + p_f V_B - T_f S_B - \Sigma \mu_{if} N_{iB} \qquad (11)$$

The subsystem available energy is defined as follows

$$A \equiv E + p_f V - T_f S - \Sigma \mu_{if} N_i \qquad (12)$$

This definition allows Eq. 11 to be expressed in terms of subsystem available energies

$$B_{AUB} = A_A + A_B \qquad (13)$$

Finally, it will be shown that subsystem available energy is an extensive property; i.e., that $A_{AUB} = A_A + A_B$.

$$A_{AUB} \equiv E_{AUB} + p_f V_{AUB} - T_f S_{AUB} - \Sigma \mu_{if} N_{iAUB}$$

$$= E_B + p_f V_B - T_f S_B - \Sigma \mu_{if} N_{iA}$$

$$+ E_B + p_f V_B - T_f S_B - \Sigma \mu_{if} N_{iB}$$

$$= A_A + A_B \qquad (14)$$

Thus, for any object, its subsystem available energy equals the sum of the subsystem available energies of its parts, proving that A is extensive.

It can also be shown (2) that the subsystem available energy changes as a result of transports and/or destructions of subsystem available energy; i.e.

$$\dot{A}_{subsystem} + \dot{A}_\tau + \dot{A}_\pi \qquad (15)$$

where the transports and destructions are

$$\dot{A}_\tau = \dot{E}_\tau + p_f \dot{V}_\tau - T_f \dot{S}_\tau - \Sigma \mu_{if} \dot{N}_{if} \tag{16}$$

$$-\dot{A}_\pi = T_f \dot{S}_\pi \tag{17}$$

It is important to recognize that Eqs. 12-17 are all valid at any instant, even if p_f, T_f, μ_{if}, ... are changing.

The equations developed in this section for subsystem available energy are generally found in textbooks on thermodynamics where they are presented as the available energy of an object or system. Furthermore, the textbooks express the transport coefficients -- T_f, p_f, μ_{if}... -- as T_0, p_0, μ_{i0}...to represent the intensive properties of the ambient surrounding(s) --assumed to be in stable equilibrium. However, as shown in this article these equations really represent the contribution of a subsystem to the overall available energy of a composite system. For cases in which a subsystem is surrounded by a stable ambient subsystem that is infinite in extent then T_f, p_f, and μ_{if}...are identical to the temperature, pressure, and chemical potentials... of the ambient surroundings; in many such cases, these T_f, p_f μ_{if} can be assumed to be constant--provided that the ambient is unperturbed by other systems (not subsystems). If the (local) ambient is stable and large but affected by other systems, then T_f, p_f and μ_{if} are functions of time. If a large ambient surroundings is not stable, then T_f, p_f and μ_{if} are equal to the values of T, p, and μ at the dead state of the ambient as a system alone; these T_f, p_f and μ_{if} are functions of time, also.

The remainder of this article addresses the selection of available energy systems and subsystems as well as the choice of dead states for analyses of practical problems.

The Selection of Reference Datums for Subsystem Available Energy.

The definition of subsystem available energy, A, which is an extensive property, is crucial to practical Second Law efficiency analysis. Before a process, device, or system can be analyzed, it is necessary to ascertain (or assume or approximate) the dead states of all relevant materials and equipment. More precisely, p_f, T_f, and μ_{if} must be known for each subsystem before Eq. 12 can be employed to evaluate A of any subsystem. In theory, once the relevant system is defined, the conventional thermodynamic principles of equilibrium can be used to find the p_f, T_f, μ_{if}...,once the possible variations (3,4) are (assumed and) prescribed. In practice, the establishment of these properties with equilibrium principles involves elaborate search methods (8,9).

In most instances the determination of p_f, T_f, μ_{if}...is more straightforward, so that equilibrium calculations need not be made.

In any case, though, the first step in the analysis is to establish "the relevant composite system". That is, a process or device or plant (with its load) cannot be analyzed in isolation from the rest of the "universe". The Second Law analysis requires consideration of the composite of all sub-systems with which the process sybsystem must interact in order to accomplish the desired purpose of the plant. Thus, for example, the relevant composite system for an automotive engine plant would consist of (at least) the engine hardware, the fuel, the confined coolant, and ambient air (used for both cooling and for combustion).

Once the relevant composite system is defined, the determination of T_f, p_f, μ_{if},...can be tackled.

Stable Reference Environments. When one of the relevant subsystems is at complete stable equilibrium and is very large compared to all of the other subsystems together, it can be called a stable ambient environment. Denote its p, T, μ by p_0, T_0, μ_{i0}. Then $T_f = T_0$ for all subsystems.

Furthermore, for all materials which are not precluded from pressure equilibrium with the large system (that is, volume may be freely exchanged), $p_f = p_0$. When a material is prevented from attaining pressure equilibrium with the environment, because it is confined in an envelope (flexible or inflexible), then p_f equals $p(T_0, v_0')$ where v_0' is the specific volume of the material when it and its confining envelope are in their dead state with the environment. Similar comments hold for μ_{if}. If substance i of a material can interact with its components in the ambient environment, then $\mu_{if} = \mu_{i0}$. Otherwise $\mu_{if} = \mu_i(T_0, p_f, x_{if})$ where the x_{if} are the mole fraction at the dead state of the material.

An example of an instance where p_f equals $p(T_0 v_0')$ would be the case of the H_2O "sealed" within the equipment of a power cycle; another example would be refrigerant (say NH_3) confined inside the components of a refrigeration cycle. The refrigerant would also be an example of a case where μ_{if} differs from the value for the components in the surrounding environment; for the refrigerant, $\mu_{if} = g(T_0, p_f) = g(T_0, p(T_0, v_0'))$.

At an instant of time a large ambient subsystem may be at a stable equilibrium state, but the state may change with time. For example, the cooling water for a power cycle may be supplied from a large lake which, at a given instant is (more or less) uniform in temperature and composition. However, the temperature may vary significantly from season to season (as a consequence of uncontrollable influences of other systems, from outside the composite of lake and power plant). Theoretically, the

efficiency analysis of the power cycle would utilize the instantaneous lake temperature for T_f. Then, to analyze the annual
performance, it would be necessary to integrate the instantaneous
results over the year. However, in practice good results can
be obtained by making just a few analyses, say one for each
season, and then weighing the results--or even with just one
analysis, employing an appropriate annual averate T_0. Thus,
the integration of instantaneous results (or the summing of
incremental results) can be avoided when (i) the variations of
the T_f, p_f and μ_{if} are relatively slow and small, and (ii)
if the loads on the system are uncorrelated with T_f, p_f and μ_{if}.
    A notable counter-example, where the analysis of the system
must be instantaneous, is the case of an air-conditioning system
(10,11). The relevant ambient surroundings is the outdoor air.
Both (i) the temperature T_0 and the humidity (and hence the
μ_i--for the N_2 and O_2 as well as for the H_2O!) change rapidly
and substantially. And (ii) the load on the air-conditioning
system is strongly dependent upon outdoor temperature and
humidity.
    On the other hand, for the analysis of a chemical plant
(e.g.,12), the variations of ambient conditions may be negligible. If the largest contributions to the available energies
of materials are their chemical availabilities (12,13), then
variations of T_0 may be inconsequential. In fact, the usage
of $T_0 = 77^OF = 25^OC$ is often justifiable--for the sake of the
convenience of thermochemical property calculations--even if the
average outdoor temperature is somewhat different.

Metastable and Unstable Ambient Environments. In some
instances, an ambient environment might be at a steady state but
not at a completely stable equilibrium state. For example (8),
nitrogen, N_2, is not in stable equilibrium with the crust of
the earth and the seas. A more stable configuration of nitrogen
is in nitrates. (Conceivably, the standard composition of air
in the environment is maintained in a steady metastable state
by intrusion from an ever larger system, such as the sun and/or
the earth's magnetic field; i.e., seemingly possible variations
(3) are prevented by unrecognized influences.) Then the appro-
priate practical choice for T_f, p_f, and μ_{if} are T_0, p_0, and μ_{i0}--
the steady values in the local ambient environment.
    With this selection of T_f, p_f, and μ_{if}, the total ΣA_j,
summed over all the subsystems j besides the ambient environment,
does not represent the absolute system available energy. Rather

$$B = \Sigma A_j + B_0 \tag{18}$$

where B_0 is the system available energy, of the composite of the
ambient environment and all other subsystems, when the other
subsystems have their T, p, and μ_i reduced to the ambient values

(or to $p(T_0,V_0')$, etc.). Insofar as the other subsystems are very small compared to the ambient, B_0 is essentially the absolute available energy of the environment alone. It follows that for a subsystem

$$E_i + p_0 V_i - T_0 S_i - \Sigma \mu_{i0} = E_i + p_f V_i - T_f S_i - \Sigma \mu_{if} N_i \qquad (19)$$

where p_f, T_f, and μ_{if} are the values at the stable dead state of the composite system and p_0, T_0, and μ_{i0} are those at the pseudo dead state--at the steady state ultimately achieved by communicating with the ambient environment, which is not at stable equilibrium.

The reason that the appropriate practical choice for T_f, p_f, μ_{if} are the steady values in the metastable local ambient environment is that the available energy B_0 is not accessible practically--as long as the metastability cannot be overcome with practical means. If the metastability can be overcome, then, of course, the proper selection for T_f, p_f, μ_{if} would be the final values in the ultimate dead state of the ambient surroundings.

In some instances an ambient environment might be unstable. For example, a power plant may interact with two large environments, such as the surrounding atmospheric air (used for combustion and for dispersion of exhaust gases) and a large body of water (used for cooling). Together the two ambients might not be at equilibrium--because the air is at a different temperature from the water and/or is not saturated with water. Theoretically, the composite of the two environments has non-zero system available energy B_0. If there are no realistically practical means for obtaining B_0, then a reasonable selection of T_f, p_f, μ_{if} would be to make different choices depending upon the process or device being analyzed. For example, the analysis of the power cycle should use T_{water} for T_f in the calculation of A_i's. Whereas, the analysis of the combustion process should use T_{air} for T_f and $\mu_{H_2O,air}$ for $\mu_{H_2O,f}$.

This procedure is appropriate inasmuch as

$$B_{system} = A_{cy} + A_{co} + B_{0,(cyUw)U(coUa)}$$

where $A_{cy} = B_{cyUw}$ is the subsystem available energy of the cycle (cy) relative to the water (w), $A_{co} = B_{coUa}$ is the subsystem available energy of the combustion subsystem (co) relative to the air (a), and $B_{0,(cyUw)U(coUa)}$ is the system available energy of the overall composite. The crucial point is that $B_{0(cyUw)U(coUa)}$ is practically equal to B_{wUa}. If B_{wUa} is not realistically available in practice, then it is reasonable to use $B_{system} = A_{cy} + A_{co}$.

There is yet another type of unstable environment which needs to be considered, namely, an environment which is made unstable by influences from the plant. The guiding principles for selecting a dead state under these circumstances will be illustrated via an example. Consider a power plant which disperses SO_2 into the environment. The environment containing SO_2 is unstable; the SO_2 could react spontaneously with H_2O in the environment to achieve a more stable configuration of the sulfur--in sulfurous and/or sulfuric acid, for example. Still, neither of these acids lacks potential to cause change (i.e., neither is inert).

Theoretically, the determination of the dead state would require the search methods referred to earlier, applying the thermodynamic principles of chemical equilibrium mentioned earlier. The application of these principles requires precise specification of the relevant subsystems, of their initial chemical constitution (and of the possible variations). In practice, this information may not be accessible.

Suppose, for example, that $CaCO_3$ (i.e., limestone-- relatively abundant) is available in the environment, then the SO_2 can combine with the $CaCO_3$ (and other components of the environment) to produce gypsum, $CaSO_4 \cdot 2H_2O$, which is very inert. There is a calculable (12,13) available energy attainable from this reaction.

Two questions arise: (1) Is there another calcium-bearing compound available in the environment which could combine with the SO_2 (and other components) to yield gypsum, and which could yield more available energy in the process? (2) Is there another sulfur-bearing compound besides gypsum that would be even more inert (which means that more available energy could be obtained, say upon reducing the gypsum thereto)? These questions could never be answered definitively; the search for these other compounds could continue indefinitely. In practice, a reasonable procedure for establishing an appropriate stable configuration of an element is by (a) making a more or less quick study of chemistry textbooks and/or reference books to ascertain what compounds, bearing the element, are inert, and (b) what compounds are conceivable environmental components which could react with the element to produce the various inert compounds. Then, considering the different combinations of reactants and products, determine those which yield the maximum available energy. Ahrendts (8), and Fan and Shieh (9), have made exhaustive analyses in this manner. For example, they find that more available energy is obtained from sulfur bearing compounds if they are reduced to gypsum with $Ca(NO_3)_2$ than with $CaCO_3$. However, $Ca(NO_3)_2$ is rare in the environment, whereas $CaCO_3$ is relatively abundant. Which is the better choice? The answer lies in the definition of "the relevant system" If it includes both the Calcium Nitrate and the Calcium Carbonate, then clearly the nitrate is the appropriate choice. If it

90   THERMODYNAMICS: SECOND LAW ANALYSIS

contains only the carbonate, then a choice must be made: Either
(1) select the carbonate as the appropriate environmental component for reducing sulfur, or (2) Re-define "the relevant system",
enlarging it to include the nitrate. If nitrate is available
nearby then (2) is the proper practical choice; if it exists
only in geographically remote locations, then (1) is appropriate.
Generally, the choice is not clearcut. What is "nearby" and
what is "remote"? That is, when would it be worth expanding
"the relevant system"? Only if the available energy obtained
with the nitrate is substantially larger than with the carbonate.
Actually, it is only slightly larger, and in this case the carbonate, which is much more abundant is clearly the better choice
than the nitrate (except in peculiar locales where the nitrate
is readily available in the environment).

     Still, there may be circumstances there the carbonate may
not be an appropriate practical choice, because it may be
scarce and/or expensive. If this is the case, then the search
must be continued.

     Often, H_2O could be used. However, consider the prospect
of dispersion of SO_2 into a desert environment; it is
conceivable that the dead state of the sulfur would need to be
taken as the SO_2, at its partial pressure in the air in the
immediate vicinity of the power plant--or even at its partial
pressure in the exhaust gases--and at ambient temperature.
From the practical standpoint, such a choice would be justified,
assuming that there would be no realistic means for utilizing
any available energy the SO_2 would have relative to desert
environment.

     On the other hand, if H_2O is available in the environment,
then the SO_2 could for example, be converted to acid. But
the acid would not be inert, unless dilute. The extent to
which it could be diluted would depend upon the amount of H_2O
in "the relevant system". If the plant is near to the sea, then
the concentration of sulfate ions in the sea would dictate the
dead state configuration of the sulfur (i.e., the extent of
dilution of $SO_4^=$); see (14) for the listing of available energy
values of various elements relative to standard sea water.

Closure.

     This article has provided the guidelines for practical
selection of reference datums for available energy. The
application of these guidelines is illustrated by the various
Second Law analyses presented (or referred to) in this volume.
     Several authors have proposed chemical reference datums
for several elements and compounds, in various environmental
circumstances; e.g., se (8,9,12,13,14).
     Although it is often intimated to the contrary, a crucial
point is this: In engineering practice the selection of an
appropriate reference datum must take into account the plant

being analyzed and its surroundings. There is no one "theoretically correct" reference environment. The practical analysis of
a plant should reflect its real circumstances. Nevertheless,
the various reference environments for chemicals which have
been proposed are helpful to the practitioner, as prospective
choices of reference datums--or as guides to the formulation of
an appropriate reference datum.

Two final points need to be made:
(1) Difficulties with the selection of chemical reference
    datum should not be allowed to be an obstacle to the
    application of Second Law analysis. The practitioner
    is encouraged to rely on judgment, and then proceed.
    Substantial errors will result only if the judgment is
    very poor. (Furthermore, judgment is improved by
    experience.)
(2) If the practitioner is willing to forego the informa-
    tion given by the absolute values of available energy
    flows, but would be satisfied with the evaluation of
    available energy consumptions only, then
    (a) Available energy balances can be used to evaluate
        the consumptions without even selecting a dead
        state for chemical available energy. Because,
        the consumption will always equal differences in
        available energies, so that the dead state values
        will cancel. In any case, though, a dead tempera-
        ture, T_f, is needed.
    (b) In fact, if only consumptions, A_c, are desired,
        then it is unnecessary to use available energy
        balances. Instead, entropy balances can be used,
        since $A_c = T_f \dot{S}_\pi$. Generally, entropy balances
        are simpler to use, since they involve fewer
        thermochemical property calculations.
    However, the evaluation of the efficiency of a process
    requires absolute values of the available energy--
    for example, the available energy content of the
    fuel. Similarly, Second Law costing depends upon
    evaluation of absolute values of available energies.
    Thus, in order to make efficiency analyses, or to do
    costing, at least an approximation to the reference
    datum must be made.

Literature Cited

1. Gaggioli, R.A., "Principles of Thermodynamics," this
   volume.

2. Wepfer, W.J., "Application of the Second Law to the
   Analysis and Design of Energy Systems," Ph.D. Dissertation, U. of Wisconsin, Madison, 1979.

3. Keenan, J.H. Thermodynamics. Wiley: New York, 1941.

4. Hatsopolous, G.N., and Keenan, J.H. Principles of General Thermodynamics. Wiley: New York, 1965.

5. Obert, E.F. Thermodynamics. McGraw-Hill: New York, 1948.

6. Obert, E.F. Concepts of Thermodynamics. McGraw-Hill: New York, 1960.

7. Obert, E.F., and Gaggioli, R.A. Thermodynamics. 2nd ed., McGraw-Hill: New York, 1963.

8. Ahrendts, J., "Die Exergie Chemisch Reaktionsfahiger Systeme," VDI-Forschungsheft 579, VDI-Verlag, Dusseldorf, 1977.

9. Fan, L.T. and Shieh, J.H., "Thermodynamically-based Analysis and Synthesis of Chemical Process Systems," presented at the U.S.D.O.E. Workship on the Second Law of Thermodynamics, George Washington U., Aug. 15-17, 1979; to be published in Energy: The International Journal.

10. Wepfer, W.J., Gaggioli, R.A., Obert, E.F. "Proper Evaluation of Available Energy for HVAC." Trans. A.S.H.R.A.E., 85, 1 (1979)

11. Gaggioli, R.A., Wepfer, W.J., and Elkouh, A.F. "Available Energy Analysis for HVAC, I. Inefficiencies in a Dual-Duct System." Energy Conservation and Building Heating and Air-conditioning Systems, A.S.M.E. Symposium Volume, H00116, 1978, 1-20.

12. Gaggioli, R.A., and Petit, P.J. "Second Law Analysis for Pinpointing the True Inefficiencies in Fuel Conversion Systems" or "Use the Second Law First." Chemtech, 7, 8 (1977), 496-506.

13. Rodriguez, L., "Calculation of Available Energy Quantities," this volume.

14. Szargut, J. and Dziedziniewicz, C., "Energie utilisable des substances chimiques inorganiques," Entropie, 40, 14-23, 1971.

RECEIVED October 17, 1979.

Available-Energy Utilization in the United States

GORDON M. REISTAD

Department of Mechanical Engineering, Oregon State University,
Corvallis, OR 97331

Other works in this symposium series present the basics of
available energy concepts and the associated Second-Law efficiency,
the merits of their use in the proper evaluation of energy systems,
and a number of specific examples of application. This paper has
the objective of viewing the overall energy flow in the U.S. from
an available energy viewpoint as well as illustrating the relative
second law efficiencies of a significant number of energy conversion and utilization systems.

Energy and Available Energy Flow in the U.S.

Figure 1 illustrates a typical energy flow diagram for the
U.S., in 1970, of the type which was first introduced by Cook(1)
and rapidly picked up and used with modifications by many others,
(see (2,3) for instance). In viewing such diagrams, this author
(and others) realized that while this type of diagram provided
some needed overview of our energy system, the diagram also did
not tell the whole story, and in fact was quite misleading on
several points. Such a diagram using available energy rather
than energy would show the truer thermodynamic picture. Figures
2 and 3 present available-energy flow diagrams for the U.S. for
the years 1970 and 1975, from Reistad (4) and Thermo Electron (5)
respectively.

Let us first consider a comparison of Figures 1 and 2, and
later consider the differences between Figures 2 and 3. Figures
1 and 2 are drawn on the same format to illustrate the difference
between energy and available-energy analyses. The energy flow
diagram illustrates that for every unit of energy that is utilized
approximately one unit of energy is wasted. The available-energy
flow diagram of Figure 2 shows a much different picture of our
technology: for each unit of available energy consumed in end uses,
greater than three units of available energy is wasted. To put
it another way, Figure 2 reveals that our level of technology in
energy conversion and utilization is roughly one-half of that indicated by the usual energy "picture" as shown in Figure 1. From
the brighter side, Figure 2 reveals that there is substantially

0–8412–0541–8/80/47–122–093$05.00/0
© 1980 American Chemical Society

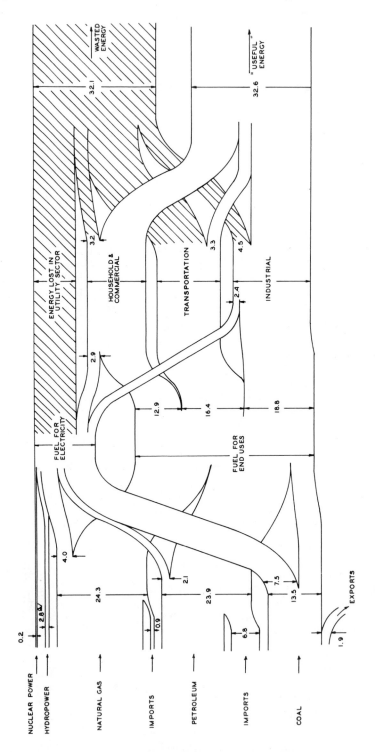

Figure 1. Flow of energy in the U.S. (4). All values reported are 10^{15} Btu $[1.055(10^{18})]$. Footnote a: the energy value for hydropower is reported in the usual manner of coal equivalent.

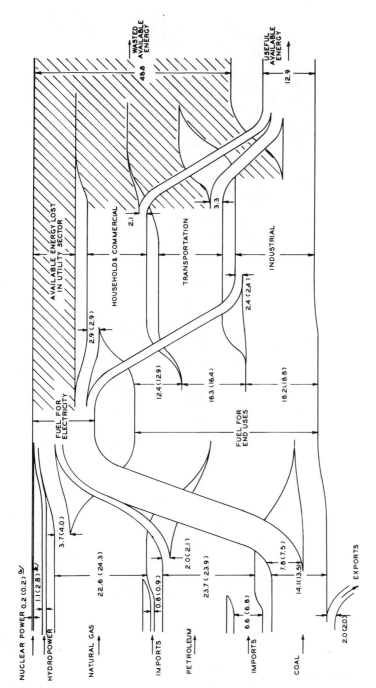

Figure 2. Flow of available energy in the U.S. (All values reported are 10^{15} Btu [1.055 (10^{18})] (4).) Footnote a: all values in ( ) are energy, other values reported are available energy. Footnote b: the energy value for hydropower is reported in the usual manner of coal equivalent; actual energy is 1.12×10^{15} Btu.

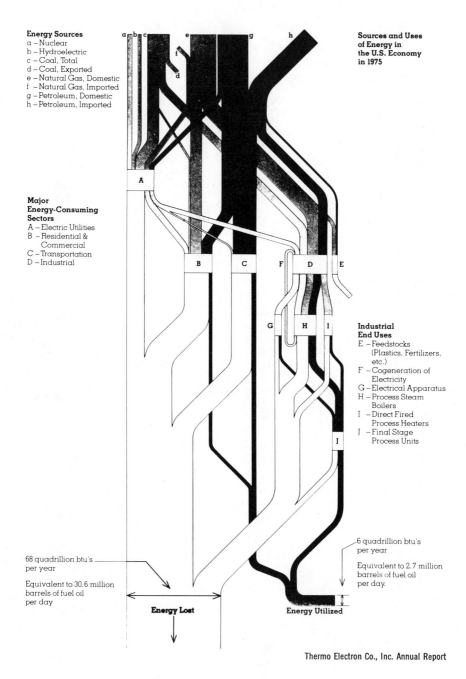

Energy Sources
a – Nuclear
b – Hydroelectric
c – Coal, Total
d – Coal, Exported
e – Natural Gas, Domestic
f – Natural Gas, Imported
g – Petroleum, Domestic
h – Petroleum, Imported

**Sources and Uses
of Energy in
the U.S. Economy
in 1975**

**Major
Energy-Consuming
Sectors**
A – Electric Utilities
B – Residential &
      Commercial
C – Transportation
D – Industrial

**Industrial
End Uses**
E – Feedstocks
      (Plastics, Fertilizers,
      etc.)
F – Cogeneration of
      Electricity
G – Electrical Apparatus
H – Process Steam
      Boilers
I – Direct Fired
      Process Heaters
J – Final Stage
      Process Units

68 quadrillion btu's
per year

Equivalent to 30.6 million
barrels of fuel oil
per day

Energy Lost

6 quadrillion btu's
per year

Equivalent to 2.7 million
barrels of fuel oil
per day.

Energy Utilized

Thermo Electron Co., Inc. Annual Report

Figure 3.   Flow of available energy in the U.S., 1975 (5)

greater room for improvement in our technology than Figure 1
indicates is possible.

When considering which segment of our economy is most in
need of improvement, magnitude of waste may be a good indication.
The energy flow diagram indicates that the electrical generation
and transmission sector and the transportation sector were the
largest contributors to the wasting of energy, and that both were
substantially less efficient than either the industrial or household and commercial segments. Figure 2 on the other hand, shows
each of the four segments of our economy contributing roughly
equal shares to the waste of available energy. Here, the transportation and household and commercial categories are indicated
as the least efficient with second law efficiency values roughly
half that for electrical generation and industrial.

The flow diagram of Figure 2 shows second-law efficiencies
for the utility sector and the three usual end-use sectors,
residential and commerical, transportation, and industrial. These
values were the author's best approximation after a modest effort
to obtain fairly exact values in 1973. Some of the values are
known to be quite good, while others are necessarily rough approximations since, for example, in the industrial sector there is
such a variety of uses in the thousands of industrial operations
that it is impossible, without a monumental effort, to determine
accurate values. The methods and data used to estimate the
second law efficiency values of each sector for Figure 2 are.
described in Appendix A.

Figures 2 and 3 are both available-energy flow diagrams for
the U.S. They are for the years 1970 and 1975 respectively and
were constructed by different authors. The five years difference
in time for which the charts are constructed is too short for
major changes in the overall performance to have occurred in the
total energy conversion system; however, it is noted that Figure 3
illustrates an even bleaker picture of our energy conversion
technology than Figure 2 does, with the overall second law efficiency from Figure 3 being 8.1% while that in Figure 2 is 17.3%.
The reason for the difference between these two diagrams is not
the change in our technology between 1970 and 1975, but rather a
difference in analysis technique by the authors of the respective
diagrams. Before considering these differences in analysis in
detail, notice that comparisons of Figure 1 with either Figure 2
or Figure 3 lead to many of the same conclusions:

-    The energy conversion technology in the U.S. is
     substantially poorer than indicated by a first law
     analysis, and consequently there is substantial room
     for improvement.
-    It is not the electricity generation and automotive
     sectors alone that are the major causes of waste as
     indicated by the first law analysis. Rather, all
     sectors have waste of about the same order (~ 40%
     variation from smallest to largest) and electricity
     generation is the smallest of these.

-    The household and commercial sector is not the most
     efficient as illustrated in Figure 1, but instead it is
     the least efficient according to both Figures 2 and 3.
The one main difference between Figures 2 and 3 in addition to
different overall efficiencies indicated previously is that
Figure 2 shows the industrial sector to be the most efficient
while Figure 3 shows the industrial sector to be only slightly
more efficient than the transportation sector and less efficient
than the electricity generation sector. This discrepancy is
attributable to the different individual process second-law
efficiencies that were used to construct the two figures. There
are two main reasons why the values of the second law efficiencies
are different for the two charts:

1.   In the industrial sector there has developed a substan-
      tial amount of information regarding the temperature at
      which processing takes place. This information was not
      available at the time of the construction of Figure 2.
2.   As discussed in another paper in this symposium (6),
      specification of a first or second law efficiency
      requires specifying the task to be performed. Quite
      often the task is relative and not absolute. Figures 2
      and 3 in a number of instances have considered different
      tasks for a specific process. The biggest difference
      is in the industrial sector where the assumption for
      construction of Figure 2 was that a certain amount of
      heating (or steam) was required at a specified tempera-
      ture level, and the task then was this heating at that
      temperature level. However, in the work on which Figure
      3 is based, the analysis is carried a step further and,
      for example in the chemical and allied products indus-
      tries, the task was not simply the heating, but the
      change in chemical composition. This is illustrated in
      Figure 3 where H represents the raising of the steam,
      while J represents the final stage process of applying
      the steam to provide the resulting chemical available
      energy change.

Energy Conversion and Utilization Performance

     It is informative to consider the individual process efficien-
cies to see which ones are low, contributing substantially to the
waste illustrated in Figures 2 and 3. This section considers the
performance of a number of general energy conversion devices as
well as a number of specific applications within certain sectors
on both a first law and second law basis.
     Presented in Table I are first law efficiency, η_I, and
second law efficiency, η_{II}, values of a wide variety of processes.
To give a single value for the performance of systems of varying
size and design that operate at varying load and under changing
conditions is a drastic simplification. For this reason, the

TABLE I.   FIRST AND SECOND LAW EFFICIENCIES OF ENERGY CONVERSION AND UTILIZING SYSTEMS.   (Largely from (4))

System	n_I (%)	n_{II} (%)
Large Electric Generator	98[99-96]	98
Hydraulic Turbine	90[93-80]	90
Large Electric Motor (over 5 hp)	90[95-85]	90
Small Electric Motor (1-5 hp)	70[85-75]	70
Storage Battery	80[90-75][b/]	80
Fuel Cell Power Plant System (methane fueled)	45[55-20]	48
Large Steam Boiler	91[92-88]	49
Diesel Engine	36[44-30]	36
Home Gas Furnace[c/]	70[85-60]	13
Home Oil Furnace[c/]	60[70-45]	11
Automobile Engine	25[30-17]	25
Steam Electric Generating Plant (coal fired)	38[42-33]	36
Home Electric Heat Pump (cop=3.5)[2.0-4.5][c/]	---	60(23)[d/]
Home Electric Resistance Heater[c/]	100[d/]	17(6.5)[d/]
Home Electric Hot Water Heater[e/]	93[d/]	16(6.2)[d/]
Home Gas Water Heater[e/]	62[70-30]	12
Home Electric Refrigerator (cop=0.9)[f/]	---	9.6(3.6)[d/]
Electric Air Conditioner (cop=2.5)[2.0-4.0][g/]	---	17(6.5)[d/]
Home Electric Cooking[h/]	80	22.5(8.5)[d/]
Home Electric Clothes Drying[i/]	50	9.5(3.6)[d/]
Home Gas Clothes Dryer[i/]	50	10.3
Incandescent Lamp[j/]	5	4.8
Fluorescent Lamp[j/]	20	19.5

a/   The dead state temperature, T_0, is assumed to be 490°R (272°K) for space heating and 555°R (308°K) dry bulb with 525°R (292°K) wet bulb for air conditioning. For other processes the exact value of T_0 is usually not critical but a value of 510°R (283°K) will be used.
b/   Usual efficiency reported, really an effectiveness.
c/   The required temperature of heating is assumed to be 590°R (328°K).
d/   The value in parentheses includes the inefficiency of the electrical generation and transmission assuming an efficiency of 38%.
e/   The water is heated to 610°R (339°K).
f/   This seemingly low cop value is reported as average by ASHRAE Guide and Data Book, Equipment Vol. 1969, p. 467. The effectiveness value is calculated assuming 1/3 load from freezer box at 460°R (256°K) and 2/3 load from cooler box at 495°R (275°K).
g/   The required temperature of cooling is assumed to be 520°R (280°K).
h/   The required temperature of cooking is assumed to be 710°R (394°K). The efficiency value is from Reference 9.
i/   The required temperature of heating is assumed to be 630°R (350°K). The efficiency value is from Reference 9.
j/   Reference 7.
--   Reference 7,8,9,10,11 were the main source of efficiency values.

thermal efficiency column of Table I contains two types of entries:
The entry in brackets represents the range of efficiency values
that might be expected for the various newly built systems operating under typical conditions. The single-valued entry merely
represents a point within the range; if a single-valued entry is
not shown with a corresponding range, either (1) the particular
system usually operates within a quite small range around the
point given and/or (2) range data for the system was unable to be
found so a representative value is presented, with the reference
indicated. The exact value of the efficiency of each system as
reported in Table I is not critical, since here the main issues
are the difference between η_I and η_{II} values and the approximate
values.

Notice for several processes η_I and η_{II} are equal. Of these
processes, the electric generator, hydraulic turbine, and electric
motors, all have energy inputs and outputs that are entirely
available; hence $\eta_I = \eta_{II}$. The storage battery is shown with
equal efficiency and effectiveness values because the usual
efficiency reported for batteries is based on the Gibbs Free
Energy change for the cell, and this is really a second law
efficiency.

For the fuel cell-power plant system, η_{II} is slightly higher
than η_I, for the diesel and automobile engine they are equal, and
for the steam-electric generating plant η_{II} is slightly less than
η_I. Here, the energy inputs are all by chemical reaction and the
available energy release of the chemical reaction is closely
equal to the energy release. Since the energy outputs of these
processes are either electricity or mechanical energy that is
entirely available, η_{II} must be closely equal to η_I, being slightly
greater than, equal to, or slightly less than it depending on the
energy to available energy ratios of the particular fuel.

The large fossil-fuel-fired steam boiler, gas furnace, oil
furnace, electric resistance heater, electric hot water heater,
gas water heater, electric cooking, electric clothes drying and
gas clothes drying all have dramatically lower η_{II} values than η_I
values. The cause for the relatively low η_{II} is that while the
input to these processes is closely entirely available, the output energy quantities are heat transfers which are far from
being entirely available (and, which in an ideal system would
require substantially less available energy than energy to supply
them). Notice that the steam boiler for a power plant which has
a relatively high grade heat transfer has the highest η_{II} for this
group. On the other hand, the home furnaces for example supply
a quite low-grade heat transfer and have a low η_{II}.

Notice the values in parenthesis (Table I) for the electric
powered processes; these values indicate the effect of the overall
process with the electricity being generated at an efficiency of
38%.

Both fluorescent as well as incandescent lamps have η_{II} and
η_I values that are closely equal. The energy output of the lamp

in the form of light is at a very high level and is approximately
entirely available.

For the home electric heat pump, refrigerator, and air conditioner, only an η_{II} value has been shown since the energy efficiency fails to be applicable in these cases since η_I might indicate a value greater than 100 percent, an indication that further
reveals the inadequacy of η_I to properly reflect either performance
or potential for improvement.

Table I reveals that a number of processes, which are widespread, have quite low η_{II} when compared with the rest of the processes shown. These processes are inferior technology and work is
needed to improve or replace them. Considering those with an η_{II}
less than 20 percent, the following are indicated: home gas
furnace, home oil furnace, home electric resistance heater, home
electric hot water heater, home gas water heater, electric refrigerator, electric air conditioner, electric clothes dryer, gas
clothes dryer, incandescent lamp and fluorescent lamp. Notice
that all of these processes are the utilization processes which,
with the exception of lighting, have not received the R&D support
at the level our other systems have.

It must be emphasized that the values in Table I are, for the
most part, those η_{II} values used to construct Figure 2 rather
than Figure 3. (Appendix A describes in detail how the values from
Table I were used, with sector energy use data, to develop Figure
2.) To re-delineate some of the difference between the evaluations that resulted in Figures 2 and 3, let us consider home
electric hot water heating. The 16% (6.2% when considering the
electricity generation) entry in Table I has as an assumption
that the task is to provide heat transfer at 610°R (339 K) in a
surroundings where the dead state is 510°R (283 K). The analysis
behind Figure 3, however, assumed that the basic task was to heat
water from about 510°R to about 610°R and for a first law efficiency of 93% (same as Table I) this would result in an η_{II} of
8.6% (3.3% when considering the electricity generation). Certainly
the latter calculation is closer to the actual desired product
and for that reason is to be preferred. However, even this value
is not precise because of the time variation of the source water
temperature as well as the dead state temperature. The important
point, evident from both evaluations, is that these low temperature heating processes are presently being satisfied in a very
inefficient way and there is substantial room for improvement.

Additional insight into the reasons for specific values of
η_{II} for each of the devices listed in Table I can be obtained by
considering several aspects of each device. Table II illustrates
three devices which are broken down into processes in which the
major available energy losses or irreversibilities occur: (i)
the boiler for high pressure steam, (ii) the furnace for comfort
heating and (iii) a large reciprocating internal combustion
engine. With the nomenclature of this table the availability
supplied ($\dot{A}_s$) is (i) consumed ($\dot{A}_c$) in carrying out the process

TABLE II.   AVAILABLE ENERGY PRODUCTIONS, LOSSES AND CONSUMPTIONS AS A
            FRACTION OF AVAILABLE ENERGY SUPPLY FOR SEVERAL DEVICES (12)*

Device	$\dot{A}_c/\dot{A}_s$	$\dot{A}_\ell/\dot{A}_s, (\dot{E}_\ell/\dot{E}_s)$	$\eta_{II} = \dot{A}_p/\dot{A}_s, (\eta_I = \dot{E}_p/\dot{E}_s)$
Boiler (high-pressure steam)	0.45	0.05,(0.1)	0.5,(0.9)
-Combustion	0.3		
-Heat Transfer	0.15		
-Chemical Effluent		0.04,(0.01)	
-Thermal Effluent		0.01,(0.09)	
Furnace (comfort heating)	0.65	0.25,(0.4)	0.1,(0.6)
-Combustion	0.3		
-Heat Transfer	0.35		
-Chemical Effluent		0.05,(0.05)	
-Thermal Effluent		0.20,(0.35)	
Engine (large, reciprocating)	0.35	0.25,(0.6)	0.4,(0.4)
-Combustion	0.20		
-Heat Transfer	0.15	(0.3)	
-Exhaust		0.25,(0.3)	

*Energy losses ($\dot{E}_\ell$) and productions ($\dot{E}_p$) are also shown as a fraction of energy supply ($\dot{E}_s$) but note that there is no energy consumption since energy must be conserved.

TABLE III.   PERFORMANCE OF TYPICAL OVERALL SYSTEMS AND THE INDUSTRIAL SUBSECTOR *(12)

	$\dot{A}_c/\dot{A}_s$	$\dot{A}_\ell/\dot{A}_s, (\dot{E}_\ell/\dot{E}_s)$	$\eta_{II} = \dot{A}_p/\dot{A}_s, (\eta_I = \dot{E}_p/\dot{E}_s)$
Fossil-fired Steam Power Plant	0.55	0.06, (0.59)	0.4, (0.41)
- Boiler	0.45	0.05, (0.09)	
- Turbines	0.05		
- Condenser and heaters	0.05	0.01, (0.5)	
Co-generating Power Plant (410 kw electricity, 1130 kw steam)	0.65	0.05	0.3, (0.75)
Co-generating Power Plant (10,000 kw elec, 17,000 kw steam)	0.62	0.05	0.33, (0.75)
Equivalent, conventional	0.62	0.10	0.28
- 10,000 kw electric power	18/30	2/30	10/30
- 17,000 kw 50 psig boiler	44/70	8/70	18/70, (55/70)
All-electric Total Energy	0.65	0.06, (0.67)	0.28, (0.33)
- Power production & trans.	0.55	0.06, (0.67)	0.33
- Heat pump	0.05		
Fossil-fired Total Energy	0.42	0.3, (0.4)	0.28, (0.6)
- Engine	0.37	0.3	
- Heating & cooling	0.05		
Equivalent, conventional	0.60	0.19	0.21, (0.5)
- Electricity	30/55	3/55	18/55
- Heating & cooling	30/45	11/45	3/45, (30/45)
Heating and Air-conditioning			
- Air-conditioning	0.85	0.1	0.04, (>1.0)
- Refrigerating unit	0.5	0.06	
- Compressor	0.15		
- Condenser	0.15	0.06	
- Expansion valve	0.06		
- Evaporator	0.15		
- Air-handling unit	0.3	0.04	
- Distribution	0.05		
- Heating	0.8	0.1	0.09, (0.6)
- Boiler	0.5	0.1	
- Air-handling unit	0.25		
- Distribution	0.05		
Coal Gasification (Koppers-Totzek)	0.3	0.06	0.65
- Coal preparation	0.06	0.005	
- Gasifier	0.15	0.02	
- O$_2$ production	0.04	0.003	
- Heat recovery	0.02		
- Gas cleanup	0.035	0.03	

* The entries are by-and-large for systems operating under optimal conditions - design loads, careful maintenance. For example, the effect of varying loads on the year-around performance of an air-conditioning system would reduce the overall effectiveness to 0.02 or below. For heating, the reduction, not so drastic, is about 0.07.

(internal irreversibilities), (ii) lost from the device ($\dot{A}_\ell$) or
(iii) the availability of the product of the device ($\dot{A}_p$). Note
that in the case of the boiler for example, 45% of the supplied
availability is consumed in the boiler as irreversibilities due to
combustion (30%) and heat transfer through a large temperature
difference (15%). The losses from the system because of thermal
losses and the chemical availability of the effluent streams
amount to 5% of the availability supplied. So for this case, the
largest degradations are in combustion and heat transfer through
a temperature difference. In the comfort heating furnace, we see
however, that three degradations are quite large, that due to combustion, that due to heat transfer through a temperature difference and that due to thermal effluent losses. These are 30%, 35%
and 20% of the supply availability respectively, illustrating that
now the heat transfer irreversibility (unaccountable by the first
law) is dominating here.

It is also interesting to consider various types of overall
systems made up of combinations of the devices in Table I, rather
than just the specific devices or unit systems. Table III shows
the performance of several overall energy systems. (Note that
the η_{II} values presented in Table III reflect to a large degree the
task definitions which resulted in Figure 3 and therefore are somewhat lower than those for Table I which were used for Figure 2.)

Conclusions

The quite often used energy flow diagrams similar to Figure 1
are quite misleading in two important aspects. First, they imply
substantial waste in the wrong sectors, pointing the finger of
blame regarding our energy problem in the wrong direction. Secondly, they imply a technology state of a substantially higher level
than we presently have; that is, they show "efficiencies" that are
much higher than properly evaluated efficiencies would be for a
substantial number of processes. On the other hand, the available
energy flow diagrams show the true picture and can be used to gain
useful insight into our overall energy problems.

The available energy diagram illustrates that our overall
level of technology is quite low, with an overall η_{II} of less than
10% with the latest figures (Figure 3). This point at first
glance seems to be a negative aspect in our future, but in fact
it is a very positive one. Since the present efficiencies are so
low there is a lot of room for improvement in our conversion systems and consequently a lot of room for reducing our energy consumption through improving the performance of energy conversion
and industrial processing (especially chemical) systems.

Literature Cited

1.   Cook, E., "The Flow of Energy in an Industrial Society,"
     Scientific American, Sept. 1971, 225 (135).

2. AEC Report on Energy Research and Development, Dec. 1, 1973.
3. "U.S. Energy: Where we Get It and Where It Goes," National
    Geographic, Nov. 1972, 142 (662-663).
4. Reistad, G.M., "Available Energy Conversion and Utilization
    in the United States," ASME J. of Engrg. for Power, July 1975,
    97 (429-434).
5. Thermo Electron Co., Inc. Annual Report, July 1977.
6. Gyftopoulos, E.P. and Widmer, T.F., "Availability Analysis,"
    This Symposium Volume.
7. Summers, C.M., "The Conversion of Energy," Scientific American,
    Sept. 1971, 225 (149).
8. Jackson, Henry M., Conservation of Energy, Committee on
    Interior and Insular Affairs, U.S. Senate, Washington, D.C.,
    1972.
9. Mark's Mechanical Engineers Handbook, McGraw-Hill, New York,
    1964.
10. "Apollo Spurred Commercial Fuel Cell," Aviation Week and
    Space Technology, Jan. 1, 1973.
11. Lueckel, W.J., and Eklund, L.G., "Fuel Cells in Decentralized
    Power Generation," Pratt and Whitney, Apr. 1972.
12. Gaggioli, R.A. and Petit, P.J., "Second Law Analysis for Pin-
    pointing the True Inefficiencies in Fuel Conversion Systems,"
    ACS Symposium Series, 1976, 21 (56-75).

APPENDIX A (largely from (4))

Second Law Efficiency Values Used in Figure 2, by Sector.

Utility

The raw energy to the utility sector is reported in reference
(A-1) as shown in Table A-1. The average heat rates reported
yield first law efficiencies of 31 percent and 31.6 percent for
nuclear and fossil-fueled plants respectively. The efficiency of
hydrogeneration is estimated as 80 percent for the water turbine
and 99 percent for the electric generator for an overall value of
79 percent. Hydropower and nuclear energy (The evaluation of the
availability of nuclear fuels has not been completely resolved at
this time; this author feels that nuclear energy is substantially
available and for purposes here will assume it to be entirely
available.) are both entirely available so the second law efficiency is the same as the efficiency for these type plants.
Adjustment of the efficiency of fossil-fueled plants by the
availability-energy ratios of the fuels yields an η_{II} of 31.6
percent.

Combination of these individual values by their relative
use results in an overall η_{II} value of 35 percent for the utility
sector (Note, here hydropower input is not on a coal equivalent
basis.).

Residential and Commercial

Energy consumption in the residential portion of this section is reported in Table A-2 according to end use for purchased electrical energy and direct use of all fuels. The data of Table A-2 and the individual η_{II} values of Table 1 result in overall η_{II} values of the residential portion of this sector for electrical use of 21.9 percent and for direct fuel use of 12.5 percent. Because of a lack of data, the η_{II} of the commercial use is assumed to be the same as the residential portion. Consequently, the η_{II} of the residential and commercial sector is evaluated by combining the above η_{II} values with the relative use of electrical and direct-fuel, in this sector, as given on Figure 2 resulting in an overall value of 13.7 percent.

Transportation

In this sector, the usual rated-load efficiencies of the individual prime movers are well established and there is fairly definitive data separating the various categories. The main uncertainty here concerns estimation of proper part-load efficiency for usual operation.

Table A-3 presents the split of the transportation sector into the categories of cars, trucks, aircraft, and railroads plus other (A-2). Included in Table A-3 are the usually accepted

Table A-1. Energy for Utility Sector

Nuclear Power	0.2 (10^{15})Btu [2.1 (10^{17})J][a]
Hydro Power	2.8 (10^{15})Btu [29.5 (10^{17})J][b]
Natural Gas	4.0 (10^{15})Btu [42.2 (10^{17})j][c]
Oil	2.1 (10^{15})Btu [22.2 (10^{17})J][c]
Coal	7.5 (10^{15})Btu [79.1 (10^{17})J][c]

[a] Average nuclear power heat rate for 1970 was 11,000 Btu/kw.
[b] Reported as coal equivalent.
[c] Average fossil fuel heat rate for 1970 was 10,800 Btu/kw.

Table A-2.   Energy Use in the Residential Sector,[a] By Type of Use.

Use Category	Energy Use	
	Electrical (%)	All Fuel (%)
Space Heating	14.9	79.5
Water Heating	14.4	15.0
Cooking	6.1	4.5
Clothes Drying	3.6	1.0
Refrigeration	17.6	---
Air Conditioning	12.1	---
Other[b]	31.3	---
	100.0	100.0

a/   Estimated 1970 values from 1968 use figures and average annual
     growth rates for period 1960-1968. (A-1).
b/   Assumed to be two-thirds illumination, 90% of which is in-
     candescent, and one-third miscellaneous mechanical drives of
     80% efficiency.

Table A-3.   Energy Use in the Transportation Sector by Mode (1970
             Values)

Mode	% of Sector Total	Principal Fuel	Rated-Load Efficiency	Estimated Operating Efficiency
Cars	55	Gasoline	25%	15%
Trucks	21	Gasoline & Diesel	30%	25%
Aircraft	7.5	Jet Fuel	35%	28%
RR & Others	16.5	Diesel	35%	28%

Table A-4.   Consumption of Energy in the Industrial Sector Accord-
             ing to End-Use. (1970 Values) (8)

Type of Use	Percent of Total U.S. Consumption
Heating (Space and Industrial Use)	14
Process (Steam)	17
Electric Drives	8

Table A-5.   Relative Energy Consumption in the Industrial Sector
             By Industry[a/]

Industry	Relative Energy Consumption	
	Raw Fuel	Electrical
Primary Metals	20.7%	23.0%
Chemical & Allied Products	17.1%	29.0%
Petroleum Refineries	13.4%	4.0%
Food	5.1%	6.0%
Paper	5.3%	5.0%
Stone	4.9%	5.0%
All Other	33.5%	28.0%
	100.0%	100.0%

a/   Data from Reference A-1 (1970 Values).

Table A-6.   Steam Generation in the Industrial Sector (A-3)

Industry	Percentage of Total Process Steam Usage	Steam Pressures and Distribution	
		Pressure Range (psig)	Distribution (%)
Chemicals & allied products	39	450-1000	3
		200-450	15
		100-200	53
		$\leq$100	29
Petroleum refining & related industries	22	150-600	20
		$\leq$150	80
Paper & allied products	18	100-200	71
		$\leq$100	29
Food & kindred products	13	50-100	10
Other industries	8	$\leq$ 50	90

rated-load efficiencies as well as this author's estimate of
operating efficiency.   These values lead to an overall η_I of 20
percent and with the availability-energy ratios of the fuels,
yield an overall η_{II} of 20 percent.

Industrial

The industrial sector is the most complicated for determining an overall n_I because of the myriad uses of energy. The problem of determining an overall n_{II} is even more complicated since for any heating process the required temperature of heating as well as the amount of heating must be known.

The split of the energy consumption of the industrial sector has been reported in broad categories of use as shown in Table A-4.

Table A-5 shows the relative industrial use according to industry types. To determine the overall n_{II}, it appeared easiest, to the author, to separate electrical and raw fuel uses. Of the total 2.4 (10^{15}) Btu [2.5 (10^{18})J] of electricity going to the industrial sector, the 8 percent of total U.S. energy usage reported in the foregoing for mechanical drives amounts to 1.65 (10^{15}) Btu [1.7 (10^{18})J]. Assuming that 90 percent of the electricity consumed in the primary metal industry is for purposes other than drives, i.e., heating, then 0.5 (10^{15}) Btu [0.53 (10^{18})J] of electricity is consumed for heating at an estimated n_I of 70 percent for heating at an assumed average temperature of 2600°F. Of the additional 0.25 (10^{15})Btu [0.26 (10^{18})J] of electricity it is assumed that 67 percent is used in industrial heating at 90 percent efficiency and an average temperature of 500°F while 33 percent is used for space heating.

Similar assumptions are made concerning the raw fuels consumption. The primary metals consume 3.9 (10^{15}) Btu [4.1 (10^{18})J] of raw fuels for industrial heating at an estimated n_I of 50 percent and an average temperature of 2600°F. The remainder of the 14 percent of total U.S. energy use that is designated for industrial heating or space heating is 2.66 (10^{15}) Btu [2.8 (10^{18})J] of which 67 percent is assumed to be industrial heating at an efficiency of 60 percent and an average temperature of 500°F with 33 percent designated to be space heating.

The additional raw fuel usage by industry amounts to 17 percent of total U.S. usage and is for generating process steam. The relative amounts and types of steam generated are presented in Table A-6.

These assumptions result in an overall n_{II} of 36 percent for the industrial sector.

Literature Cited

A-1 The Potential for Energy Conservation, A staff study, Office of Emergency Preparedness, Oct. 1972.
A-2 "The Energy Crises: Alternatives for Transportation," Automotive Engineering, Vol. 81, No. 3.
A-3 Miller, A.J., et al., "Use of Steam-Electric Power Plants to Provide Thermal Energy to Urban Areas," ORNL-HUD-14, Oak Ridge National Laboratory, Jan. 1971.

RECEIVED December 14, 1979.

Thermodynamic Balance and Analysis of a Synthesis Gas and Ammonia Plant

HELMUT CREMER

Fichtner Consulting Engineers, Boxgraben 79, 5100 Aachen,
Federal Republic of Germany

It is not intended to discuss the specific problems of the
synthetic ammonia production but to localize those areas of an
industrial chemical process which cause consumptions. The analysis comprises the synthesis gas plant using methane in a reforming process and the ammonia production. The use of methane is an
arbitrary choice. The analysis of the coupled processes is necessary to give a realistic impression and to allow a qualified judg-
ing of the units and systems.

The total plant discussed in this paper is not self-sufficient
whereas independent operation is preferred in practice; this is
rather influenced by service requirements than by thermodynamic
considerations.

The Process

The simplified flowsheet of the coupled synthesis gas and
ammonia plant is shown in Fig. 1. The streams are numbered from
1 to 44 and are listed in Table I and Table II;Table I refers to
the reactant and product streams especially. The units of the
process may be identified by Table III.

The flowsheet and data refer to a plant with an ammonia production of 1,000 tons per day and can be found in detail in chemical engineering textbooks. All data are conservative. Some
minor discrepancies in the mass streams result from the data cards
but do not influence the results significantly.

The upper part of Fig. 1 shows the synthesis gas plant which
is fed from the right-hand side with methane (stream 1) and air
(stream 2) for a combustion process to match the heat requirements
of the synthesis gas process. The combustion process delivers the
exhaust gas (stream 3). The synthesis gas is produced by methane,
water vapor and air (streams 4, 5, 6) in a primary and secondary
reformer and a converter (units REF1, REF2 and CON). The raw
gas (stream 28) passes the gas conditioning (SEP1) which has been
detailed in Fig. 3 and the synthesis gas (stream 29) enters the
ammonia plant shown in the lower part of Fig. 1. The ammonia

0–8412–0541–8/80/47–122–111$05.00/0
© 1980 American Chemical Society

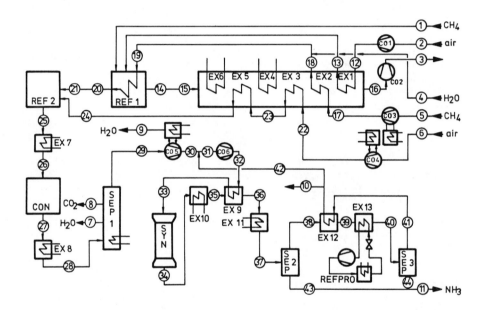

Figure 1.   Synthesis gas and ammonia plant (simplified flowsheet)

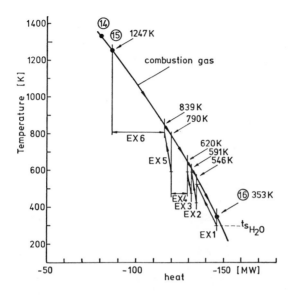

Figure 2.   T/Q̇ diagram (temperature/heat flow) cooling of combustion gas

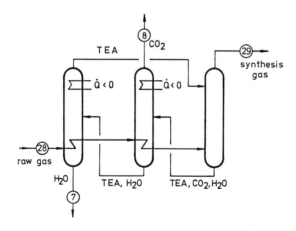

*Figure 3.   Detailed scheme of SEP1 (compare with Figure 1). Synthesis gas con-
ditioning with TEA; total heat removal—65.83 mol wt.*

Table I. Reactants and Products (compare with Figure 1) (1,000 t/d NH₃ ≙ 680 mol/sec NH₃)

Stream	Pressure bar	Temperature k	Flow rate mole/s	mole fraction %								
				N_2	O_2	Ar	CH_4	CO	CO_2	H_2O	H_2	NH_3
Combustion process:												
1	1.1	293	170				100					
2	1.0	293	1,780	78	21	1						
3	1.0	359	1,960	70.8	2.2	1.0			8.7	17.3		
Synthesis gas plant:												
4	32.0	823	1.530							100		
5	1.0	293	340				100					
6	1.0	293	478	78	21	1						
7	1.0	293	1,069							100		
8	1.0	293	338						97,6	2.4		
Ammonia plant:												
9	1	298	2							100		
10	230	293	144	22		3.3	7.1				65.7	1.9
11	230	269	680									100

synthesis is characterized by the recirculating gas stream (stream 42) which is 2.6 times the fresh gas stream (stream 30). Both streams are mixed after the high pressure compressor (C05). Within the ammonia plant there is a refrigerating process (REFPRO) to allow for the ammonia separation by condensation.

Parameters and Assumptions

The total plant is operated under steady state conditions. The operating conditions and the data of the streams are given in Tables I to III. The process has been simplified to some extent; this is necessary for a comprehensive representation. (E.g., most units are adiabatic; but heat loss is considered at the hot gas piping (14 → 15 and 20 → 21) after the primary reformer (REF1).)

The thermodynamic data of Barin and Knacke (1) and of Davies (2) have been used.

Thermodynamics

The well known notations of the First and Second Laws are used to do an energy and availability balance and analysis of the process.

Under steady state conditions the availability stream $\dot{A}$ of an enthalpy stream $\dot{H}$ and the availability stream $\dot{A}_{\dot{Q}}$ of a heat flux $\dot{Q}$ are defined by

$$\dot{A} = \dot{H} - \dot{H}_0 + (\dot{E} - \dot{E}_0)_{kin,pot} - T_0 (\dot{S} - \dot{S}_0) \qquad (1)$$

$$\dot{A}_{\dot{Q}} \quad \frac{T - T_0}{T} \quad \dot{Q} \qquad (2)$$

(T: absolute temperature; $\dot{S}$: entropy stream; $\dot{E}$: energy stream; Index 0: ambient conditions; Index kin: kinetic (energy); Index pot: potential (energy)).

The availability consumption $\dot{A}_C$ is according to

$$\dot{A}_C = T_0 \cdot \dot{S}_{irr} \qquad (3)$$

and the entropy production $\dot{S}_{irr}$ due to irreversibilities follows from the Second Law

$$\dot{S} = \Sigma \frac{\dot{Q}}{T} + \dot{S}_{irr} \qquad (4)$$

where the first term on the right-hand side refers to the heat exchange.

The availability consumption $\dot{A}_C$ may be calculated as well by means of an availability balance for a balancing region; Bošnjakovíc (3) emphasized that the selection of the control volumes should not be abused to discredit a unit and favor another.

Table II. Streams (Compare with Figure 1)

No.: Stream number $\dot{n}$: Flow rate
p : Pressure $\dot{H}$: Enthalpy stream
T : Temperature $\dot{S}$: Entropy stream

No.	p Bar	T K	$\dot{n}$ mole/s		$\dot{H}$ MW	$\dot{S}$ MW/K
1	1.1	293	CH_4	170	-12.695	0.0315
2	1.0	293	sum	1,780	- 0.0002	0.3340
			N_2	1,388.4		
			O_2	373.8		
			Ar	17.8		
3	1.0	359	sum	1,960	-145.700	0.3972
			N_2	1,388.4		
			O_2	43.8		
			Ar	17.8		
			CO_2	170		
			H_2O	340		
4	32.0	823	H_2O	1,530	-342.094	0.3004
5	1.0	293	CH_4	340	- 25.391	0.0632
6	1.0	293	sum	478	- 0.0001	0.0897
			N_2	372.8		
			O_2	100.4		
			Ar	4.8		

Table II. Continued

No.	p bar	T K	$\dot{n}$ mole/s		$\dot{H}$ MW	$\dot{S}$ MW/K
7	1.0	293	H_2O	1,069	− 306.056	0.0736
8	1.0	293	sum	338	− 129.974	0.0703
			CO_2	330		
			H_2O	8		
9	1.0	298	H_2O	2		
10	230	293	sum	144	− 0.910	0.0155
			N_2	31.7		
			Ar	4.8		
			CH_4	10.2		
			H_2	94.6		
			NH_3	2.7		
11	230	265	NH_3	680	− 47.390	0.0635
12	1.05	299	sum	1,780	+ 0.303	0.3490
			compare 2			
13	1.03	523	sum	1,780	+ 11.914	0.3804
			compare 2			
14	1.0	1,331	sum	1,960	− 80.240	0.4843
			compare 3			
15	1.0	1,247	sum	1,960	− 86.335	0.4796
			compare 3			
16	0.96	353	sum	1,960	− 146.065	0.3993
			compare 3			
17	33	423	CH_4	340	− 23.779	0.0581

Table II. Continued

No.	p bar	T K	$\dot{n}$ mole/s		$\dot{H}$ MW	$\dot{S}$ MW/K
18	32	593	CH_4	340	− 21.055	0.06357
19	32	766	sum	1,870	− 363.149	0.3719
			CH_4	340		
			H_2O	1,530		
20	30	1,073	sum	2,391	− 283.696	0.4636
			CH_4	78.9		
			CO	112.4		
			CO_2	148.2		
			H_2O	1,123.8		
			H_2	927.7		
21	30	1,048	sum	2,391	− 285.996	0.4614
			compare 20			
22	34	473	sum	478	+ 2.482	0.0858
			compare 6			
23	33	593	sum	478	+ 4.211	0.0906
			compare 6			
24	30	823	sum	478	+ 7.681	0.0952
			compare 6			
25	29	1,303	sum	2,926	− 278.555	0.5997
			N_2	371.6		
			Ar	5.9		
			CO	196.0		
			CO_2	143.4		
			H_2O	1,249.4		
			H_2	959.7		
26	28.6	607	sum	2,926	− 351.857	0.5202
			compare 25			

Table II. Continued

No.	p bar	T K	$\dot{n}$ mole/s		$\dot{H}$ MW	$\dot{S}$ MW/K
27	28	673	sum	2,926	− 351.692	0.5237
			N_2	371.6		
			Ar	5.9		
			CO	26.3		
			CO_2	313.1		
			H_2O	1,079.7		
			H_2	1,129.4		
28	27.8	473	sum	2,926	− 371.507	0.4888
			compare 27			
29	26	303	sum	1,509	− 1.303	0.1789
			N_2	372.7		
			Ar	4.5		
			CH_4	10.6		
			H_2O	3.0		
			H_2	1,118.2		
30	230	303	sum	1,507	− 0.575	0.1601
			N_2	373.7		
			Ar	4.5		
			CH_4	10.6		
			H_2	1,118.2		
31	230	297	sum	5,543	− 25.459	0.5948
			N_2	1,263.8		
			Ar	138.6		
			CH_4	293.8		
			H_2	3,774.8		
			NH_3	72.1		
32	250	313	sum	5,543	− 22.902	0.5987
			compare 31			

Table II. *Continued*

NO.	p bar	T K	$\dot{n}$ mole/s		$\dot{H}$ MW	$\dot{S}$ MW/K
33	250	473	sum	5,543	+ 3.414	0.6618
			compare 31			
34	230	687	sum	4,760	+ 2.329	0.6800
			N_2	894.9		
			Ar	133.3		
			CH_4	290.4		
			H_2	2,684.6		
			NH_3	756.8		
35	230	503	sum	4,760	− 26.965	0.6307
			compare 34			
36	230	345	sum	4,760	− 53.282	0.5670
			compare 34			
37	230	303	sum	4,760	− 63.134	0.5358
			compare 34			
38	230	303	sum	4,437	− 41.575	0.9026
			N_2	896.3		
			Ar	133.1		
			CH_4	288.4		
			H_2	2,684.4		
			NH_3	434.8		
39	230	263	sum	4,437	− 46.481	0.4947
			compare 38			
40	230	253	sum	4,437	− 56.449	0.4550
			compare 38			

Table II.   Continued

No.	p bar	T K	$\dot{n}$ mole/s		$\dot{H}$ MW	$\dot{S}$ MW/K
41	230	253	sum	4,080	− 30.618	0.4251
			N_2	897.6		
			Ar	134.6		
			CH_4	289.7		
			H_2	2,680.6		
			NH_3	77.5		
42	230	293	sum	3,936	− 24.884	0.4231
			N_2	865.9		
			Ar	129.9		
			CH_4	279.5		
			H_2	2,586.0		
			NH_3	74.8		
43	230	303	NH_3	323	− 21.559	0.0332
44	230	253	NH_3	357	− 25.831	0.0299

TABLE III.

UNITS

(compare Fig. 1)

$\dot{A}_{ci}$:   Consumption availability in unit i

Unit	$\dot{A}_{ci}$ MW	$\dot{A}_{ci}/\Sigma\dot{A}_{ci}$ %	Remarks
CO1	4.46	2.98	compressor for combustion air; adiabatic; 1-stage; efficiency 0.7; power 0.3 MW
CO2	0.63	0.42	blower for gas outlet; compare CO1; power 0.32 MW
CO3	1.86	1.24	methane compressor; three stages and intermediate cooling; efficiency 0.65; power 5.07 MW; cooling -3.38 MW
CO4	3.83	2.56	process air compressor; compare CO3; power 7.51 MW; cooling -5.0 MW
CO5	9.02	6.02	high pressure synthesis gas compressor; intermediate and final cooling; efficiency 0.6; removal of condensed water; power 15.37 MW; cooling -15.37 MW
CO6	1.14	0.76	recirculating compressor; power 2.56 MW; adiabatic
EX1	2.46	1.64	heat exchanger; preheating of combustion air; heat flow 11.61 MW; compare Fig. 2
EX2	0.04	0.03	preheating of methane for combustion; heat flow 2.72 MW; compare Fig. 2
EX3	0.42	0.28	preheating of process air; heat flow 1.73 MW; compare Fig. 2
EX4	2.50	1.67	heat flow 10.22 MW; compare Fig. 2; this heat is not used in the process and still available e.g. for steam generation. Availability consumption results from the entropy change of the combustion gas to be cooled and an assumed heat transfer with an average temperature difference of 200 K between the thermodynamic mean

TABLE III (cont'd.)

Unit	$\dot{A}_{ci}$ MW	$\dot{A}_{ci}/\Sigma\dot{A}_{ci}$ %	Remarks
			temperatures of the combustion gas and the medium to be heated
EX5	0.29	0.19	preheating of process air; heat flow 3.47 MW; compare Fig. 2
EX6	3.51	2.34	heat flow 29.97 MW; compare Fig. 2; parameters as for EX4
EX7	7.03	4.69	heat flow 73.30 MW; heat still available; parameters as for EX4
EX8	5.67	3.78	heat flow 19.82 MW; heat still available; parameters as for EX4
EX9	0.16	0.11	preheating of synthesis gas by internal heat recovery; heat flow 19.49 MW
EX10	3.35	2.24	steam generation; heat flow 29.29 MW; arbitrary choice of parameters for the steam: complete evaporization at 50 bar including preheating from 325 K; 1 kg steam/1 kg NH_3; minimum temperature difference 30 K
EX11	0.55	0.37	cooling with ambient air and partial condensation of ammonia; heat flow 9.85 MW
EX12	1.73	1.15	preheating of recirculating gas; heat flow 4.91 MW
EX13 incl. REF-PRO	2.91	1.94	cooling (9.97 MW) by means of a refrigerating process; arbitrary choice of parameters: NH_3-refrigerator, low pressure 1.32 bar, pressure ratio 10.8, compression power 4.77 MW, adiabatic throttling
REF1	48.90	32.64	primary reformer; adiabatic; internal heat exchange 79.45 MW; equilibrium reaction on the combustion side; oxidation ratio about 1.1; reforming conditions known by measurements
REF2	13.09	8.74	secondary reformer; heat loss 0.24 MW
CON	1.04	0.69	CO-converter

TABLE III (cont'd.)

Unit	$\dot{A}_{ci}$ MW	$\dot{A}_{ci}/\Sigma\dot{A}_{ci}$ %	Remarks
SEP1	16.37	10.93	separation region for synthesis gas conditioning; Fig. 3 shows a detailed flowscheme; it has been recalculated that the heat content of the raw gas covers the heat requirements of the separation and washing agent (tri-ethane-amine) recycle; total heat removal to surrounding 65.83 MW
SEP2 SEP3	– –	– –	catchpots for liquified ammonia; without availability consumption because of zero pressure drop; compare preceding assumptions
SYN	6.51	4.35	high-pressure ammonia synthesis reactor; heat loss 1.09 MW; the reactor is assumed to produce the total pressure drop of the ammonia plant i.e. from stream 32 to 42; the actual conversion is 46.3% of the real gas equilibrium at 673 K and 240 bar
MIX-ING 4+18 =19	2.39	1.60	adiabatic mixing of preheated process methane and steam in front of the primary reformer
MIX-ING 30+42 =31	3.48	2.32	adiabatic mixing of synthesis gas and recirculating gas in the ammonia plant
MIX-ING 43+44 =11	0.13	0.09	adiabatic mixing of liquid ammonia with different temperatures
PIP-INT 14→15	4.69	3.13	heat loss at high temperature 6.1 MW
PIP-ING 20→21	1.65	1.10	heat loss 2.3 MW
	149.81	100	

Chemical processes represent multi-component systems for which the equilibrium state of the surroundings has to be defined, in order to evaluate the availability of chemical mixtures (4, 5). However, if only availability consumptions are to be evaluated, the availability itself need not be evaluated, and the specification of the reference environment can be avoided. Entropy balances can be used in lieu of availability balances, and the calculations are thereby simplified.

Results

The availability consumptions of the units are given in Table 3; the sum is about 150 MW for the 1,000 t/d ammonia plant. This figure does not include the losses of the exhaust gas (stream 3) and of the flash gas (stream 10) which stabilizes the inert gas concentration of the recirculating gas. Extensive use has been made of internal heat recovery (compare Fig. 2); but there is heat still available in the process (heat exchangers EX4, 6, 7, 8, 10, 11)--the use of which has not been optimized. The heat may be used for steam generation. A representative procedure-- used in this paper--for these heat exchangers is described for unit EX 4 in Table 3.

The relative consumptions (%) of the individual units are given in Table 3 as well. The primary reformer (REF1) produced about 1/3 (32.64%) of the total consumption whereas the high-pressure ammonia reactor contributes only 4.3% though the complete pressure drop of the recirculating gas cycle has assumed to be localized in this reactor. The gas conditioning (SEP1; compare Fig. 2) has a considerable influence with about 11%.

It may be surprising that the ratio of availability consumptions

synthesis gas plant		80.7%
to	is	to
ammonia plant		19.3%

With respect to types of units the following figures result for the total plant:

reactors	46.4%
internal heat exchange and heat recovery	18.1%
compressors	14.0%
separation	10.9%
heat losses cooling to surrounding	6.6%
mixing	4.0%

The principal conclusion, then is that attempts to make the ammonia production more efficient should start with improvement of the reactors.

Editors Note:

As pointed out by Dr. Cremer, the evaluation of availability consumptions is unquestionably simplified by the use of entropy balances rather than availability balances, inasmuch as the property calculations are less extensive. Then, results like those shown in Table 3 can be determined--namely the absolute values of availability consumptions, and the value of any one consumption as a percent of the total.

In some instances, at least, it is desirable to evaluate the efficiency of a plant and/or the unit operations. For, the importance of the absolute values of the consumptions is relative, to be compared with the input of availability to the plant (or unit), or to the product output from the plant. Also, the evaluation of availability losses with effluent outputs may be important to the assessment of inefficiencies. If it is desired to evaluate the inputs and outputs, entropy calculations alone do not suffice, unfortunately. Then, the availability must be calculated ($\underline{4}$, $\underline{5}$, $\underline{6}$). For the case at hand, the effluent losses are nearly 25 MW with stream 3 and 1.5 MW with stream 10. The availability of the NH_3 product is closely 283 MW. The overall plant efficiency is therefore:

$$\eta_{II,plant} \quad \frac{\dot{A}_{product}}{\dot{A}_{feedstock}} = \frac{\dot{A}_{product}}{\dot{A}_{product} + \Sigma \dot{A}_C + \Sigma \dot{A}_L}$$

$$= \frac{283 \text{ MW}}{283 + 150 + 26.5} = 61.6\%$$

For the ammonia plant, $\Sigma \dot{A}_C$ can be found, from Table 3, to be 25.9 MW, while $\dot{A}_L = \dot{A}_{10} = 1.5$ MW

$$\eta_{II,NH_3 \text{ plant}} = \frac{283 \text{ MW}}{283 + 26 + 1.5} = 91\%$$

Since the synthesis-gas plant and ammonia plant operate in series, it follows that

$$\eta_{II,synthesis\text{-}gas \text{ plant}} = \frac{0.616}{0.911} = 67.6\%$$

In order to determine the efficiency of the various units, it would be necessary to evaluate the availability of input and/or output flows.

Even when it is desired to calculate efficiencies and losses, it is valuable to recognize that the use of entropy balances for calculating consumptions, in the manner illustrated by

Dr. Cremer, can serve to shorten the calculations.

Finally, it should be mentioned that the article by Benedict and Gyftopoulos (7) in this volume illustrates the use of entropy balances for (i) analyzing the consumptions in an air-separation process, and (ii) for Second-Law costing analyses applied to optimal economic selection of the process equipment.

Abstract

The availability consumptions in a synthesis gas and ammonia plant have been analyzed and localized; it is shown that the synthesis gas plant--due to the dominating influence of the reforming reactors--produces 4 times the consumption of the ammonia works.

Literature Cited

1. Barin, I; Knacke, O. "Thermochemical Properties of Inorganic Substances"; Heidelberg-Berlin-New York, 1977.

2. Davies, P. "Temperature-Entropy Diagram for Ammonia" in "Thermodynamic Functions of Gases"; Butterworths Scientific Publications: London, 1950.

3. Bŏsnjakovíc, F. "Technische Thermodynamic," Vol. II; Dresden and Leipzig, 1965.

4. Szargut, J.; Styrylska, T. Brenstoff-Wärme-Kraft, 1964, 16, 589-596.

5. Gaggioli, R.; Petit, P. Chemtech, 1977, 7, 496-506.

6. Rodriguez, L. (this volume).

7. Benedict, M.; Gyftopoulos, E. P. (this volume).

RECEIVED November 5, 1979.

THERMODYNAMIC EVALUATION
OF PROCESS AND ENERGY
FACILITIES: THERMOECONOMICS

Benefit–Cost of Energy Conservation

ELIAS P. GYFTOPOULOS

Massachusetts Institute of Technology, 77 Massachusetts Ave.,
Cambridge, MA 02139

THOMAS F. WIDMER

Thermo Electron Corporation, 101 First Ave., Waltham, MA 02154

There are compelling technical and economic arguments for
adopting a balanced approach to the energy problem, one that gives
equal consideration to increased supply and to effective end-use
of energy (1–11). Historically, there has been a tendency to
focus upon energy sources, with less attention given to improving
the efficiency of processes and devices which utilize energy for
performing various tasks. To the extent that it has been introduced to the recent energy debate, conservation has too often been
equated with curtailment (e.g., lowered thermostats, reduced speed
limits, etc.) rather than improved end-use efficiency.

A supply-dominated approach was economically sound in an era
of abundant and inexpensive energy resources. Today, however, we
can no longer look forward to steadily expanding supplies of
energy at ever-declining real cost. In fact, we have seen in recent years a sharp reversal of the long-term decline (in real
terms) of energy prices. The outlook is for even higher cost in
the future, particularly for those energy forms where regulatory
actions have thus far restrained prices below the level of replacement cost.

Technologically and financially, there are ample opportunities for saving energy in all sectors of the economy. As shown in
Figure 1, the second law efficiency of energy usage in the United
States is estimated to be about 8%. Even the industrial sector,
our largest user of energy, operates at an overall efficiency of
only about 13%. Several studies have disclosed the potential
economic advantages of energy conservation for various industrial
processes. These studies (1) indicate that approximately 25%--
the equivalent of 4.5 million barrels of oil per day--of projected
1985 energy usage in manufacturing could be saved through conservation measures whose capital and total costs would be equal to or
less than those needed to obtain comparable amounts of new energy
supply.

The primary sources of energy inefficiency in manufacturing
are burning of high-grade fuel to produce low-temperature process
heat, loss of high-grade waste heat to the environment, and use of

0–8412–0541–8/80/47–122–131$05.00/0
© 1980 American Chemical Society

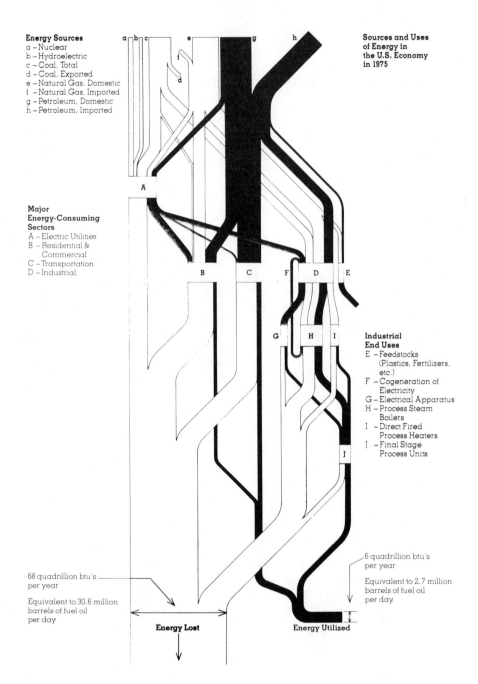

Energy Sources
a – Nuclear
b – Hydroelectric
c – Coal, Total
d – Coal, Exported
e – Natural Gas, Domestic
f – Natural Gas, Imported
g – Petroleum, Domestic
h – Petroleum, Imported

Sources and Uses of Energy in the U.S. Economy in 1975

Major Energy-Consuming Sectors
A – Electric Utilities
B – Residential & Commercial
C – Transportation
D – Industrial

Industrial End Uses
E – Feedstocks (Plastics, Fertilizers, etc.)
F – Cogeneration of Electricity
G – Electrical Apparatus
H – Process Steam Boilers
I – Direct Fired Process Heaters
J – Final Stage Process Units

68 quadrillion btu's per year

Equivalent to 30.6 million barrels of fuel oil per day

6 quadrillion btu's per year

Equivalent to 2.7 million barrels of fuel oil per day.

Energy Lost

Energy Utilized

Figure 1. Effectiveness of energy usage in the U.S. economy (based upon availability)

certain manufacturing processes having inherently large thermodynamic loss mechanisms. Inefficiencies of specific processes can be attacked only through systematic analysis of the operations particular to each industry—a task that requires intimate and detailed knowledge of the technologies involved. However, the first two sources of inefficiency are common to all manufacturers and can be reduced with technology applicable to many industries. These inefficiencies, which are found in fuel-fired kilns, furnaces, boilers, ovens, and dryers, can be reduced through various process rearrangements, including recycling of waste energy and recovery of waste energy.

Rising Replacement Costs for New Energy Supplies

The advantages of implementing a vigorous energy efficiency program in industry can best be seen by comparing the capital cost of new energy supply facilities to the cost of energy-saving equipment. Because of the diminishing store of readily recoverable fossil fuel resources, development of new energy supplies consumes an ever-increasing amount of capital for each unit of energy production capacity (3).

Apart from the Middle East, where reserves are still easily accessible, most new petroleum or natural gas production facilities (e.g., U.S. outer continental shelf, North Sea, Alaska) require anywhere from $5 billion to upwards of $10 billion for each Quad per year of equivalent fuel energy provided (1 Quad per year = 10^{15} Btu/year = 456,000 bbl per day = 33,400 Megawatts thermal). Synthetic gas and oil obtained from coal will be even more costly, probably in excess of $15 billion per annual Quad.

New coal supplies are still obtainable at a capital cost of $1.5 billion to $2.0 billion per Quad per year. However, the mining, processing, and combustion of coal is attended by serious environmental and safety problems that may ultimately limit its rate of consumption, or at least cause increases in the cost of supply. Moreover, coal does not possess the flexibility of application inherent to oil and gas. The industrial sector could undoubtedly substitute more coal for such purposes as the raising of process steam, but any major increase in our reliance upon coal will depend, for the most part, upon its greater use by electric utilities.

The capital cost is, of course, the highest when the energy is provided in the form of electricity. For every Quad per year of delivered electricity, the capital investment in facilities for fuel supply, generation, transmission, and distribution will range from $50-60 billion for coal-based systems, to over $70 billion for nuclear generation.

The sharp rise in capital cost of new energy supplies has created a large disparity between the price of energy and the cost of its replacement. From 1950 to 1970, energy prices and replacement costs declined annually in real terms at a rate of 1.7%.

Electric utilities were building larger and more efficient plants, and energy suppliers were using more sophisticated methods of discovering, producing, and distributing natural gas and oil. In those days, all forms of replacement energy, even regulated natural gas, had costs equal to or less than the price of supplied energy.

Just before the oil embargo, however, replacement costs started to climb sharply. From 1971 to mid-1973, prices rose to their real 1950 level. For the first time in our history, the replacement cost of energy sources jumped above the average price paid by consumers. The postembargo oil price increases have accelerated this trend; as Table I shows, the replacement cost for the mix of energy consumed by industry in mid-1977 exceeded the price of supplied energy by about 37%. The cost disparity in September 1977 was the most for natural gas (59%)--it would be even higher if Alaskan gas, liquified natural gas, or synthetic fuels were used as replacement fuels--and the least (0%) for coal.

Table I.
Average Price and Replacement Cost of Energy Used in Industry

Energy form	Percent of industrial use	Dollars per million Btu of delivered energy (September 1977)	
		Average spot price	Replacement cost
Coal	19.0	$1.05	$ 1.05
Petroleum	34.0	2.41	2.88
Natural Gas	34.5	2.08	3.30
Electricity	12.5	9.00	12.90
Weighted Average		$2.86	$ 3.93

For electricity, a utility's rate structure represents an average of the cost of new plant and equipment plus older, less expensive facilities. For example, a utility might charge an industrial customer at a rate of 2.6 cents per kilowatt hour (kWh), based upon an after-tax return on assets of about 8%. If this electricity were to be produced exclusively from new plant facilities, the rate would have to rise to 4.1 cents per kWh in order to realize the same return on assets. Of course, the allowable return for utilities is now usually higher than 8% (with higher prime rates, rate-setting bodies may permit returns of up to 12%). Thus, marginal costs for electricity will rise still further.

These disparities between price and replacement cost of energy cannot last, because new energy will eventually supply a large fraction of the demand, causing replacement costs to reach a new plateau. The costs of synthetic fuels, and perhaps of

electricity from either solar or breeder reactor sources, will be
at least 1.5 to 2 times the current replacement costs.

Industrial Conservation - The Effective Use of Capital

In contrast to new energy supplies, capital equipment costs
have remained almost constant in real terms. The economic potential of energy conservation using such equipment can be seen by
considering typical energy-saving measures available in certain
industrial processes (Table II). Capital costs for energy saving
are substantially lower than those for equivalent new energy supply facilities.

As previously noted, such cost-effective efficiency improvements represent an energy saving of 25% for all manufacturing
(about 10 Quads per year at 1985 projected usage). An estimated
$125 billion in capital investment would be needed to realize
these savings. This is a large investment but smaller than that
required to provide a comparable amount of energy from new supply
facilities. In fact, the capital investment needed to produce the
same 10 Quads per year of coal, petroleum, gas, and electricity
from new sources would be at least $170 billion.

Individual industries will vary substantially from the average 25% figure for savings because of wide differences in the relative efficiency of current practices compared to the most advanced technology available in each particular industry. Also,
various industries have a different mix of energy-saving opportunities with regard to fuel types (i.e., the fraction of savings
that can be achieved in coal, oil, gas, and electricity).

The lower capital cost figures for various conservation measures (relative to the investment needed for new energy supplies)
provide only a rough indication that such measures will yield attractive economic returns to the investor. Detailed payback calculations, similar to those discussed later for waste heat recovery equipment, are needed to establish economic feasibility in
each specific case.

Most manufacturers have identified methods of saving energy
that require little or no capital investment. These methods,
often simply of a housekeeping nature, have largely been imple-
mented in response to higher prices and shortages occurring in the
post-embargo era. True efficiency gains, however, require capital
investment, often of such a high level as to strain the resources
available to a particular industry. Nevertheless, it is desirable
to stimulate these conservation investments, at least up to the
point where the cost of additional conservation equals the marginal cost of energy from new fuel and electricity facilities, for
otherwise our capital resources will be misallocated and the competitive edge of our industry will be eroded.

In addition to limited capital, at the present time there are
other barriers that impede the flow of investment capital into industrial energy conservation.

Table II.
Examples of the Cost Effectiveness
of Energy Conservation Measures

Conservation Measure	Energy Saved and Type of Fuel	Capital Cost of Unit (1975)	Specific Capital Cost of Conservation Measure ($/Annual Quad)	Capital Cost of Comparable New Supply
RECUPERATORS ON STEEL REHEAT FURNACES	10^8 Btu/hr for 170 Tons/hr Steel Throughput	$480,000	$0.7 x 10^9 per Quad of fuel	10 Times Conservation
BOTTOMING CYCLES ON STEEL REHEAT FURNACES	5.0 MW electricity for 170 Tons/hr Steel Throughput	$3.4 x 10^6	$28.4 x 10^9 per Quad of electricity	2 Times Conservation
COGENERATION OF ELECTRICITY AND PROCESS STEAM IN STEEL PLANTS	40 MW electricity for 10,000 Ton per day plant	$41.4 x 10^6	$34.6 x 10^9 per Quad of electricity	1-3/4 Times Conservation
DRY QUENCHING OF COKE IN STEEL INDUSTRY	4 x 10^9 Btu/day of steam fuel plus 1.4 x 10^6 Btu/day coke (3000 Ton/day plant)	$9.3 x 10^6 (retrofit) $5.3 x 10^6 (new plant)	$4.7 x 10^9 per Quad of fuel $2.7 x 10^9 per Quad of fuel	Same as Conservation 1½ Times Conservation
REDUCED CURRENT DENSITY IN HALL PROCESS ALUMINUM REDUCTION CELLS	2160 kWh per Ton of Primary Aluminum	$350 per Annual Ton of Primary Aluminum	$48 x 10^9 per Quad of electricity	1-1/4 Times Conservation
BOTTOMING CYCLES ON CEMENT KILNS	4.7 MW electricity for 600 Ton/day plant	$2.7 x 10^6	$23.9 x 10^9 per Quad of electricity	2½ Times Conservation
AIR PREHEATERS ON ETHYLENE PLANT PROCESS HEATERS	4.8 x 10^{11} Btu/yr for 1400 Ton/day plant	$1.14 x 10^6	$2.3 x 10^9 per Quad of fuel	3 Times Conservation
STEAM GENERATION FROM AMMONIA PROCESS HEATER FLUE GAS	11 x 10^6 Btu/hr for 1000 Ton/day ammonia plant	$130,000	$1.1 x 10^9 per Quad of fuel	2 to 6 Times Conservation
FEEDWATER PREHEATERS ON STYRENE PLANT PROCESS HEATER	4 x 10^6 Btu/hr for 400 Ton/day styrene plant	$17,000	$0.6 x 10^9 per Quad of fuel	4 to 12 Times Conservation
OPTIMIZATION OF CRUDE DISTILLATION PREHEAT TRAIN	23 x 10^6 Btu/hr 100,000 bbl/day refinery	$1.0 x 10^6	$4.7 x 10^9 per Quad of fuel	1½ Times Conservation
POWER RECOVERY TURBINES ON HIGH PRESSURE REFINERY PROCESS FLOW STREAMS	11.0 MW electricity for 100,000 bbl/day refinery	$2.8 x 10^6	$8.2 x 10^9 per Quad of electricity	8 Times Conservation
INCREASED USE OF BARK AND WOOD WASTES AS FUEL IN PAPER INDUSTRY	10^9 Btu/day for 400 Ton/day mill	$1.7 x 10^6	$4.8 x 10^9 per Quad of fuel	1½ Times Conservation

- The average price of energy charged to manufacturers has not yet caught up with the rapidly increasing replacement cost of new energy supply.
- Energy users require a higher rate of return on their investment than do regulated energy producers (e.g., gas and electric utilities) who face lower risks.
- Most manufacturers demand a higher rate of return on so-called "discretionary" investments (e.g., conservation) relative to main-stream investments (e.g., expansion of capacity), because investments critical to maintaining market share must be given high priority.
- Federal, state, and local regulations restrict the generation of byproduct electricity by many manufacturers.

These factors play an important part in the decision making process of any manufacturer considering energy conservation investment. In fact, most manufacturers will not invest in conservation measures having a payback period in excess of about 2 years. An illustration of how these factors can distort an investment decision can be seen from an actual case study.

In November 1976, the manager of a plant located in a Southern state presented to his corporate management a proposal to install a 4700-kW electricity generator powered solely by waste heat. The turnkey price of this "bottoming cycle" generator was $2.7 million. Upon review by management, the proposal was rejected; the plant continues to purchase electricity from a utility. This electricity is generated by consuming the equivalent of about 188 barrels of petroleum a day. Had the company elected to purchase the waste heat recovery unit, the electricity produced would have replaced not only an equal amount of new fuel supply facilities, but also a corresponding amount of new generating capacity earmarked by the utility in its expansion plans. The items that would have been replaced cost either $5 million or $8 million for coal- or nuclear-based systems, respectively--two to three times the cost of the energy-saving proposal.

Top management of the company assessed the plant manager's proposal on the basis of 2.6 cents per kWh paid for purchased electricity. This analysis resulted in a projected after-tax return on investment (ROI) of 22%, considerably less than the 30% expected from the discretionary investments that do not increase production. Although the company's typical ROI criterion was only about 15% and its average return on assets only 10%, top management rejected the proposal.

If, however, the company had been paying for electricity the 4.1-cent replacement cost appropriate to new electricity supply, the generator would have earned a return well above 30% and management would have undoubtedly approved the project.

Payback Periods of Selected Conservation Measures

Capital spending analysis consists of evaluating the investment outlay in terms of the economic gain that it will provide.

The investment outlay is represented by the net investment required for the new equipment or change in operation, and the economic gain is represented by estimated operating cash flows generated by the investment. Different companies use different criteria for comparing outlays and gains.

A widely used method of analysis is that based upon the internal rate of return. Also known as the Discounted Cash Flow method, this approach is based on the criterion that the sum of the present value of all cash flow returns associated with a given project be equal to the initial investment outlay; namely,

$$C_o = \sum_{n=1}^{N} \frac{NCF_n}{(1 + i)^n}$$

where

C_o = initial investment outlay
NCF_n = net cash flow at year n
i = internal rate of return
n = year, 1, 2.....N
N = economic life in years

The factor $1/(1 + i)^n$ transforms each expenditure or net cash flow to its value at time zero. The net cash flow at year n is defined as the savings resulting from the investment, i.e., savings resulting from a reduction in purchased fuel or electricity, minus the costs chargeable to operation and maintenance (O&M), federal taxes, and loan payments.

$$NCF_n = S_n - OM_n - FT_n - ALP_n$$

where

S_n = savings on fuel for year n
OM_n = operation and maintenance costs for year n
FT_n = federal taxes for year n
ALP_n = annual loan payments for year n

The required rate of return varies with the company, industry, and general economic climate. An average required rate of return, based upon the weighted averages of debt and equity, value line risk factors, bond ratings, and riskless borrowing rates, can be obtained for an industry for all its investments. The industry-wide average of the after-tax rates of return calculated from these parameters for the chemical, petroleum, and paper and pulp industries, and for utilities is 15%, 16%, 15%, and 13%, respectively.

Another method of analysis consists of evaluating the gross payback period, namely the ratio of the initial investment outlay C_o for a project to the yearly pre-tax gross savings GS_1 at the time of investment ($GS_1 = S_1 - OM_1$). The gross payback period is related to the internal rate of return. Obviously, changes in tax regulations, depreciation, accounting, financing procedures, escalation in savings and costs, etc. will change directly the internal rate of return and indirectly the gross payback period.

Figure 2 shows the relationship between internal rate of return and gross payback for a specific set of tax regulations (10% investment tax credit, 50% tax rate), 6% per year increase in power, fuel, and O&M costs, and sum-of-digits depreciation for both 10-year life (for recuperators and heat exchangers) and 20-year life (for power generation equipment and boilers). We see that life has little effect on the internal rate of return for gross payback periods less than about 3 years.

Recent studies by Thermo Electron Corporation have shown that certain forms of waste heat utilization (e.g., recuperators or process steam boilers) provide a better return on investment than that obtainable from bottoming cycle generators. The optimum choice of heat recovery strategy depends, in part, upon the temperature of exhaust heat available.

As shown in Table III, the waste heat from just 3 major industries (chemicals, petroleum refining, and paper) represents over 5 Quads per year of total energy--equivalent to about 1/6 of all energy consumed by the entire industrial sector. The largest potential for high-temperature (600°-1000°F) heat recovery is in petroleum refining. The chemical industry has a large amount of waste heat at intermediate (300°-600°F) temperatures, and all three industries have large amounts in the low-temperature ($<300^\circ$F) range. Because over 70% of the waste heat is at temperatures below 300°F (with over 50% under 150°F), there is generally less opportunity for economical heat recovery in the paper and pulp industry than in either the chemical or petroleum refining industries.

Table III.
Summary of Available Waste Heat in the
Chemical, Petroleum Refining, and
Paper and Pulp Industries

Industry	Temperature Ranges				Total Amount 10^{15} Btu/yr
	$<150^\circ$F	150-300°F	300-600°F	600-1000°F	
	Percent of Total				
Chemical	7	44	44	5	2.3
Petroleum Refining	15	29	29	27	1.2
Paper and Pulp	52	20	28	0	1.6

Current trends in heat recovery include retrofitting existing process furnaces and installing new furnaces with combustion air

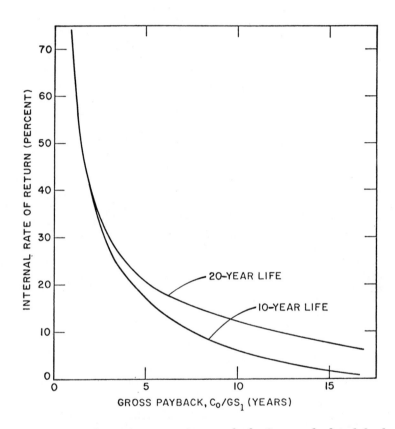

Figure 2. Internal rate of return vs. gross payback. Gross payback is defined as the ratio of capital expenditures to pretax gross savings at the time of investment (C_0/GS_1): 50% tax rate; 10% investment tax credit; S.O.D. depreciation; 6% per year increase in power, fuel, and operation and maintenance costs.

Table IV.
Comparison of Paybacks for Different
Recovery Measures

ENERGY RECOVERY METHOD	Payback (Years)	
	500°F Flue Gases	800°F Flue Gases
Electrical Generation by Bottoming Cycle		
20 x 10^6 Btu/hr waste heat	~8.0	4.7
40 x 10^6 Btu/hr waste heat	5.5	3.5
80 x 10^6 Btu/hr waste heat	3.8	3.6
Combustion Air Preheating		
20-80 x 10^6 Btu/hr waste heat	~1.0	~1.0
Steam Generation (125 psig)		
20 x 10^6 Btu/hr waste heat	1.8	0.8
40 x 10^6 Btu/hr waste heat	1.5	0.7
80 x 10^6 Btu/hr waste heat	1.2	0.4
Power Recovery from Product Stream by Means of Hydraulic Turbine	1.0	2.0

Based on: 7000 hours of operation per year, fuel at
$2.00/$10^6$ Btu, electricity at $0.03/kWh.

preheating (especially in the petroleum industry where 600°-1000°F
waste heat is available), generating process steam with waste heat
boilers, adding product stream heat exchangers, and providing
power recovery turbines in pressurized power streams.

Comparative gross payback periods for the various energy recovery approaches are shown in Table IV. For high-temperature
flue gases (800°-1000°F), process steam generation appears to be
more economical than combustion air preheating or electrical generation by bottoming cycles. At a 500°F flue gas temperature, the
reduced steam production (caused by the boiler pinch-point problem) makes combustion air preheating more economical than steam
generation. At both flue gas temperatures, the economic paybacks
between the two waste heat recovery technologies are only mar-
ginally different. On the other hand, electrical generation by
bottoming of waste flue gases (with steam or organic working fluid
Rankine cycles) is generally not competitive with either combustion air preheat or steam generation. For flue gas temperatures
above 800°F, sufficiently high-pressure steam can be generated to
make the economics of back-pressure turbines (steam expanded to
125 psig for use in process) for electrical production approach
those for combustion air preheating. In practice, the choice between combustion air preheat, steam generation, or electric generation is dictated not only by economics, but also by the specific
plant requirements for steam or electricity and by the availabil-

ity of natural gas and/or oil. For a process with few steam requirements, electrical generation may be the optimum selection.

As energy prices continue to rise towards their replacement
cost levels, industry will undoubtedly respond by increasing investment in energy-saving process equipment. Those manufacturers
who move quickly to take advantage of available conservation opportunities can expect to gain a significant competitive advantage
over their less efficient rivals.

Literature Cited

1. Hatsopoulos, G.N., Gyftopoulos, E.P., Sant, R.W., and Widmer,
    T.F., "Capital Investment to Save Energy," Harvard Business
    Review, Vol. 56, No. 2, March–April 1978.
2. Widmer, T.F. and Gyftopoulos, E.P., "Energy Conservation and a
    Healthy Economy," Technology Review, Vol. 79, No. 7, June
    1977; pp. 31–40.
3. "A Comparison of Capital Investment Requirements for Alterna-
    tive Domestic Energy Supplies," American Gas Association Re-
    port, Planning & Analysis Group, May 1978.
4. "Technical Aspects of Efficient Energy Utilization," American
    Physical Society, Summer Study, 1974.
5. "A Study of In-plant Electric Power Generation in the Chemical,
    Petroleum Refining and Paper and Pulp Industries," Thermo
    Electron Corporation Report No. TE5429-97-76.
6. "Toward a National Energy Policy, Capital Requirements," Mobil
    Oil Corporation, 1975.
7. Gyftopoulos, E.P., Lazaridis, L.J., and Widmer, T.F., "Poten-
    tial Fuel Effectiveness in Industry," report to the Energy
    Policy Project of the Ford Foundation, Ballinger Publishing
    Co., 1974.
8. Berg, C.A., "A Technical Basis for Energy Conservation,"
    Technology Review, Vol. 76, No. 4, February 1974; pp. 14–23.
9. Schipper, L., "Toward More Productive Energy Utilization,"
    Annual Review of Energy, Vol. 1, Annual Reviews, Inc., 1976;
    p. 455.
10. Alexander, T., "Industry Can Save Energy Without Stunting Its
    Growth," Fortune Magazine, May 1977.
11. "A Study of Improved Fuel Effectiveness in the Iron and Steel
    and Paper and Pulp Industries," Thermo Electron Corporation
    Report No. TE5429-71-76.

RECEIVED November 1, 1979.

Available-Energy Costing

GORDON M. REISTAD

Department of Mechanical Engineering, Oregon State University, Corvallis, OR 97331

RICHARD A. GAGGIOLI

Department of Mechanical Engineering, Marquette University, 1515 W. Wisconsin Ave., Milwaukee, WI 53233

It is the available energy content of the commonly labeled "energy" commodities, not their energy, which really represents their potential to cause changes such as to do work, heat, convert chemicals, etc. Consequently, it is the available energy of these commodities that is used up as a process proceeds, and it is the available energy, then, that is worth paying for. In economic decision making, it thus makes sense to do available energy accounting rather than the energy accounting that is quite often done.

The available-energy concept associated with the Second Law of Thermodynamics goes back to Maxwell (1) and Gibbs (2) and has had many names such as available-work, energie-utilisable, exergy, essergy, potential energy, availability . . . (The authors feel that the lack of a uniformly adopted appropriate nomenclature has been a significant factor in the resistance to widespread use of the available energy concepts.) It is ironical that, although the concept arose in English-speaking countries, and even though a few American thermodynamicists have been emphasizing its importance for some time, only in the last several years has available energy started to receive serious interest in the U.S., at least at the practical level.

It was Keenan (3) who first emphasized that cost accounting should be based on available energy, not energy. Consequently, Obert and followers (5-13) have made straightforward applications of available-energy costing in some uncomplicated cases of practical importance, while Tribus and followers (14,15,16) have made important theoretical contributions.

This paper reviews the use and advantages of using the available-energy concept for costing "energy" commodities. The basic principles which underlie available energy accounting are presented and explained. Also, various uses of such accounting are also described--in particular, applications to (1) cost accounting, (ii) design optimization, and (iii) system operation and maintenance. The aim, then, is to provide a good understanding of the principles and the realm of available energy accounting.

0–8412–0541–8/80/47–122–143$05.00/0
© 1980 American Chemical Society

Available Energy

Available energy is the thermodynamic property which measures the potential of a system to do work when restricted by the inevitable surroundings at T_0 and P_0 and $\mu_{j,0}$ . . . . Unlike energy, availability is not conserved, and any degradation, such as friction or heat transfers through a temperature difference, or chemical conversions through chemical potential differences, results in destruction of availability. The destruction of availability is often called irreversibility and is a lost ability to do work, heat, convert chemicals, etc. Derivation of the basic availability equations is straightforward (see for example (3) through (7)) and will not be repeated here. The significant results, for considerations in this paper, are embodied in the steady-flow availability balance applicable to any steady process, device or system:

{Net rate of availability supply to a process} =

    {Net rate of availability output from the process}

  +   {Net rate of availability loss in effluents}

  +   {Net rate of availability consumption by the process}

Or,

$$\Sigma P_{A,supplies} = \Sigma P_{A,outputs} + \Sigma P_{A,effluents} + \dot{A}_{consumed} \qquad (1)$$

In this symposium volume, Petit and Gaggioli (17) present equations which relate the transports, P_A, to various commodities with which availability is commonly carried, and Rodriguez and Petit (18) present means for evaluating the P_A from thermodynamic property data.

Comparing equation (1) to the steady-flow energy equation (equation (2)) illustrates the significant difference between the two properties energy and availability:

$$\Sigma P_{E,supplies} = \Sigma P_{E,outputs} + \Sigma P_{E,effluents} \qquad (2)$$

As indicated in equation (2), energy is conserved. But available energy need not be conserved and is, in fact, destroyed (consumed) in every real process, as reflected by the irreversibility, $\dot{A}_c$ in equation (1), which must always be positive.

The performance parameter of availability analysis is the second-law efficiency (The second-law efficiency has been termed effectiveness (3); but, to decrease confusion with the different "effectiveness" of heat exchangers, the term second-law efficiency is preferred.), denoted by η_{II}, and defined as

$$\eta_{II} \equiv \frac{\text{availability of desired output}}{\text{availability supplied}} \qquad (3)$$

For example, in a simple mechanical-compression refrigeration system, the refrigeration is the desired product, so the numerator in equation (3) becomes the availability of the refrigeration; the denominator, being the availability supplied to drive the system, is the power input to the compressor. To compare with the energy analysis, consider the energy or first-law efficiency η_I:

$$\eta_I \equiv \frac{\text{energy of desired output}}{\text{energy input}} \qquad (4)$$

The merit of the two performance parameters has been discussed elsewhere (3,5,8); and, for purposes here, let it suffice to recap briefly the essence of these discussions. The second-law efficiency is the performance parameter which indicates the true thermodynamic performance of the system. It is the η_{II}, not the η_I, which measures how well a device (or system) is performing, compared to the optimum possible performance. In other words, $1 - \eta_{II}$ not $1 - \eta_I$ is a measure of the potential for improvement.

The first-law efficiency is useful only for comparing systems which have like (equal grade) inputs and like (equal grade) outputs, or for comparing two models of a device.

Available Energy Accounting

Available energy accounting in "energy" systems is important in regard to a number of types of engineering-economic decision making. This section presents the rationale and basics of available-energy cost accounting. It summarizes material from references (6) and (8-16), wherein additional examples and details may be found. The types of decision making to be considered here are (i) the proper costing of different energy utilities supplied from a common system, (ii) design optimization, and (iii) component operation and maintenance strategies.

Costing Energy Utilities. In many instances, an industrial concern is engaged in the manufacture of products some of which are steam-intensive, some of which are electricity-intensive, others of which are compressed-air-intensive, and so on. In these cases, it is important for management to debit each product properly for the cost of the electricity, the steam, the compressed-air, etc., used in its manufacture. Similarly, a company which co-generates may wish to sell excess steam or electricity to an outside concern: it needs to know the correct costs of these commodities. Also, an electric utility may be distributing and selling steam or hot water. In each of these situations, a method must be devised to determine properly the costs of the "energy" flow streams. The following discussion illustrates that the "energy" flows should not be costed on the basis of their energy content, but that costing them on the basis of their available-energy content does provide a rational method.

Money Balances. The first step in costing "energy" flow streams is the application of money balances to the system of interest. The cost (C) of the product of an energy-converter equals the total expenditure made to obtain it--the fuel expense, plus the capital and other expenses:

$$C_{product} = C_{expended} \equiv C_{fuel} + C_{capital}, \text{ etc.}$$

The average unit cost, c_p, of the product is the total cost, $C_{product}$ divided by the output power: $c_p \equiv C_{prod}/P_{prod}$. Similarly, the fuel cost can be expressed as $C_{fuel} \equiv c_f P_{fuel}$ where c_f is the unit cost of fuel and P_{fuel} is the power input to the converter. One may then write

$$c_p = \frac{c_f P_{fuel}}{P_{prod}} + \frac{C_{capital}}{P_{prod}}$$

Recall that the efficiency is simply P_{prod}/P_{fuel}, so that

$$c_p = \frac{c_f}{\eta} + \frac{C_{capital}}{P_{prod}}$$

The first term on the right reflects fuel costs, and the second the capital (and other) costs.

Costing on the Basis of Energy Content: It is common to cost power flows on the basis of energy content. Such costing with energy can lead to radical errors, which lead to misconceptions, poor decisions, and misappropriations. As emphasized by Keenan[3], the proper commodity to be costed is the available energy. The following example, for costing co-generated steam and shaft power, will illustrate these claims.

When steam and electricity are co-generated, it is often critical to know how much of the costs should be attributed to each commodity. The first step in determining the unit cost of the turbine shaft work and that of the "back-pressure steam" (the exhaust steam from the turbine) is to obtain the unit cost of the high-pressure steam supplied to the turbine. This cost can be determined by applying the foregoing expression for the unit cost of product, c_p, to the boiler:

$$c_{HP\ steam} = \frac{c_{coal}}{\eta_{boiler}} + \frac{C_{boiler}}{P_{HP\ steam}}$$

In turn, the outputs of the turbine, driven by the high-pressure steam, are shaft-work and low-pressure steam. A money balance yields the total cost of these products as

$$C_{shaft\ work} + C_{LP,steam} = C_{expended} \equiv c_{HP}P_{HP} + C_{turbine}$$

$$c_{shaft}P_{shaft} + c_{LP}P_{LP} = c_{HP}P_{HP} + C_{turbine}$$

If the purpose of the steam turbine is to convert power supplied in steam to shaft power, then the shaft power should be
debited for the cost of the power <u>extracted</u> from the steam,
$[P_{HP} - P_{LP}]$, as well as for capital and other costs:

$$c_{shaft} P_{shaft} = c_{HP} [P_{HP} - P_{LP}] + C_{turbine}$$

while

$$c_{LP} = c_{HP}$$

Dividing the former equation by the shaft power

$$c_{shaft\ work} = c_{HP} \frac{P_{HP} - P_{LP}}{P_{shaft}} + \frac{C_{turbine}}{P_{shaft}}$$

Considering the co-generating power plant indicated in
Figure 1 and Table I, and with the shaft power and steam costed on
the basis of their <u>energy</u> content (i.e., the power flows will be
measured with energy), the unit cost of the high-pressure steam is
(neglecting the relatively low availability value of the water
entering the boiler)

$$c_{HP} = \frac{(\$1.60/10^6 \text{ Btu})}{0.87} \cdot 3413 \frac{\text{Btu}}{\text{kWh}}$$

$$+ \frac{0.08(\$10(10^6)/\text{yr})(3413 \text{ Btu/kWh})}{(180,000 \text{ lbm/hr})(0.7)(8760 \text{ hr/yr})(1377.8-18.1)\text{Btu/lbm}}$$

$$= 0.81\text{¢/kWh}$$

and, per pound,

$$c_{HP} = (0.81\text{¢/kWh}) (\frac{1377.8-18.1}{3413} \frac{\text{kWh}}{\text{lbm}}) = 0.32\text{¢/lbm} \ .$$

In turn

$$c_{shaft} = 0.81\text{¢/kWh} + (\frac{0.08(\$70/\text{kW})(10300 \text{ kW})/\text{yr}}{(10300 \text{ kW})(0.7)(8760 \text{ hr/yr})}) = 0.90\text{¢/kWh}$$

and

$$c_{LP} = c_{HP} = 0.81\text{¢/kWh}$$

or, on a per pound of steam basis,

$$c_{LP} = (0.81\text{¢/kWh}) \frac{1182 - 18.1}{3413} \frac{\text{kWh}}{\text{lbm}} = 0.276\text{¢/lbm}$$

If such energy-based costing is valid under the above circumstances, with back-pressure steam being delivered at 50 psig, then
it should also be appropriate when the back-pressure is 40 psig,
or 30, 20, 10, . . . Figure 2, curve A, shows the results of just

Table I. Basic Cost and Performance Data of a 10,000 kW, 50 psia
          Back Pressure Co-generating Power Plant.

Description:

Shaft work output	10,300 kW
Steam from boiler	180000 lbm/hr at 750°F, 650 psia
Boiler cost	$10(10^6)$: \$5.6$(10^4)/10^3$ (1bm/hr)
Turbine cost	$0.7(10^6)$: \$70/kW
Unit cost of coal	\$1.60/10^6 Btu
$n_{I,boiler}$	0.87
$n_{II,boiler}$	0.36
$n_{II,turbine}$	0.85
Load factor	0.7
Amortization factor	0.08

such costing. The curve indicates that steam at pressures like
0.5 psia (T = 80°F) would be worth about 0.21¢/1bm--almost as much
as 50 psia steam--even though it has virtually no usefulness for
heating. Certainly this is wrong; there would be no buyers of 0.5
psia steam at 0.21¢/1bm--at (0.21¢/1bm)/((944 - 18.1) Btu/1bm) =
\$2.26 per million Btu of _energy_. There is no logical way to decide
at which back-pressures energy costing would be appropriate and at
which pressures it would not be. Essentially, this shows that the
error is in the use of _energy_ as the measure for the power flows.

Costing on the Basis of Available Energy: This same procedure
will now be followed, but with _available energy_ to evaluate the
power flows for the case in which the co-generating power plant
exhausts 50 psia steam. For steam produced at 650 psia and super-
heated to 750°F, the available energy per pound is (in this work,
$T_0 = 510°R$, $P_0 = 1$ atm.)

$$a_f = h - T_0 s - \mu_0 = h - T_0 s - (h_0 - T_0 s_0) = 571 \text{ Btu/1bm} .$$

Thus, the average unit cost of high pressure steam leaving the
boiler is

$$c_{HP} = \frac{(\$1.60/10^6 \text{ Btu})(3413 \text{ Btu/kWh})}{0.36}$$

$$+ \frac{(0.08(\$10(10^6)))/yr)(3413 \text{ Btu/kWh})}{(18000 \text{ 1bm/hr})(0.7)(8760 \text{ hr/yr})(561 \text{ Btu/1bm})} = 1.96¢/kWh$$

The unit cost of shaft work is then given by

$$c_{shaft} = 1.96 \frac{¢}{kWh} \frac{1}{0.85}$$

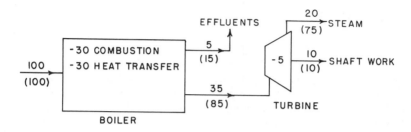

Figure 1.   *Cogeneration power plant showing available energy and energy flows at one operating condition*

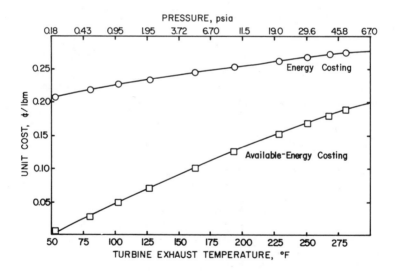

Figure 2.   *Unit steam costs as a function of steam exhaust pressure on the basis of energy and available energy*

$$+ \; \frac{(0.08(\$70/kW)(10300kW)/yr)}{(10300 \text{ kW})(0.7)(8760hr/yr)} = 2.4\cancel{c}/kWh$$

Also, $c_{LP} = c_{HP} = 1.96\cancel{c}/kWh$. The available-energy content of the steam is 561 Btu/lbm and 331 Btu/lbm, respectively, for high- and low-pressure steam; and, hence,

$$c_{HP} = (1.96\cancel{c}/kWh) \; (\frac{561}{3413} \frac{kWh}{lbm}) = 0.32\cancel{c}/lbm$$

$$c_{LP} = (1.96\cancel{c}/kWh) \; (\frac{331}{3413} \frac{kWh}{lbm}) = 0.19\cancel{c}/lbm$$

Table II compares these available-energy based costs to the energy based costs. Notice the substantial difference in the cost per pound of low-pressure and high-pressure steam when costed on the available-energy basis, in great contrast to the closely equal costs from the energy-based costing. Indeed, the high-pressure (high-temperature) steam is more valuable and would maintain a significantly higher price.

Table II. Results of Energy Costing and Available-Energy Costing of Electricity and Steam from a Co-Generating Plant. (Table I lists the basic data for the plant considered here.)

	Energy Costing	Available-Energy Costing
high pressure steam	0.32¢/lbm	0.32¢/lbm
shaft work	0.9¢/kWh	2.4¢/kWh
low pressure (50 psi) steam	0.276¢/lbm	0.19¢/lbm

The results of repeating this available-energy costing scheme, applied to the co-generation system in which the exhaust pressure is gradually lowered, are shown on Figure 2. Available-energy costing yields that the cost per pound of low-pressure steam does go to zero as its usefulness goes to zero. This is precisely the result that any rational costing scheme should provide.

This example has clearly illustrated that energy-based costing can lead to errors and that unit costs should be obtained from available-energy based costing methods.

Costing Methods. Consider an energy-converter, for example a boiler, that takes in fuel and produces a single output such as steam. A money balance on the boiler gives

$$C_{HP \; steam} = c_{HP}P_{HP} = C_{fuel} + C_{boiler}$$

Once the fuel cost and boiler costs are known, all that remains is

to choose a measure of the product and solve for the unit cost of
product. For single-output devices it makes no difference which
measure of product output is chosen; it could be mass, energy,
available energy, and so on. For example, refer to Table II and
note that the unit cost of high-pressure steam is the same for the
energy and the available-energy based costing schemes.

As in the case of the back-pressure turbine, some energy-
converters yield multiple outputs. Recall that the money balance
on the turbine of co-generation systems resulted in the expression

$$c_{LP}P_{LP} + c_{shaft}P_{shaft} = c_{HP}P_{HP} + C_{turbine}$$

Generally the right-hand side and the power flows, P_{LP} and P_{shaft},
are known. This results in one equation with two unknowns:
the unit costs c_{LP} and c_{shaft}. Another equation is needed. There
are various methods for obtaining a second equation. Which method
is appropriate is outside the province of thermodynamics, and
depends on economic considerations.

The presentation which follows is for co-generation systems
that have two or perhaps three outputs (say at two extraction
points and shaft output; or steam, compressed-air and shaft power).
The economic considerations underlying various methods for obtaining additional cost equations are discussed.

The method used in the previous example, called the extraction
method, assumes that the sole purpose of the turbine is to produce
shaft power. Therefore, the shaft work is charged for the capital
cost of the turbine and for the steam available energy used by the
turbine to produce the work. With this rationale, the additional
equation is obtained by equating the unit costs of high- and low-
pressure steam available-energy, $c_{LP} = c_{HP}$. The result is that the
shaft work bears the entire burden of the costs associated with the
turbine process and capital expense.

Insofar as the purpose of a turbine is to produce shaft work,
the extraction method is appropriate since it charges the work for
the turbine capital and inefficiency. However, if there were no
turbine, the efficiency of producing any required low-pressure
steam would be lower and, hence, its unit cost higher. So, when
both low-pressure steam and shaft work are required commodities
it is appropriate to charge this steam proportionately for the
costs attributable to the turbine. The equality method charges
the shaft work available-energy and that of the low-pressure steam
available-energy equally, for the cost of high pressure steam and
turbine capital: $c_{LP} = c_{shaft}$. Then, the money balance on the
turbine gives

$$c_{shaft} = (c_{HP}P_{HP} + C_{turbine})/(P_{LP} + P_{shaft})$$

Sometimes an industrial concern may view the shaft-work as a
by-product-- say because it produces vast amounts of required steam

and/or for historical reasons. The by-product work method assumes
that the steam delivered by the turbine would have to be produced
anyhow (say to meet process requirements), even if no power were
produced. The available-energy in steam is, therefore, costed as
if it, alone, were produced--in a hypothetical low-pressure boiler
which would deliver steam at the pressure and temperature at which
it is needed:

$$c_{LP} = \frac{c_{fuel} P_{fuel}}{P_{LP}} + \frac{C_{LP\ boiler}}{P_{LP}} = \frac{c_{fuel}}{\eta_{II,LP\ boiler}} + \frac{C_{LP\ boiler}}{P_{LP}}$$

The low-pressure boiler has a lower efficiency, η_{II}, than does a
high-pressure boiler; and, as a result, the cost c_{LP} of low-pressure
steam available-energy will be greater than for high-pressure steam.
Once this c_{LP} is known, the money balance on the turbine can be
solved for c_{shaft}. This method results in unit cost values for the
low-pressure steam that are substantially higher than the equality
or extraction methods. Conversely, the cost of shaft work will be
lower. An example of an instance when it would be proper to use
the by-product method for costing shaft work would be the following. Suppose a concern has determined that it will be adding substantial steam generation capacity. It is considering the prospects
of cogenerating power for its own use. Then, if the prospective
cost of the power, determined by the by-product method, proves to
be less than the cost of purchased power, the concern ought to co-
generate. Once the decision to co-generate is made, then it is
another matter--a different economic problem--to determine how the
steam and power should be costed in order to charge the "consumers"
within the concern properly. For examples, the extraction or
equality method might be appropriate.

Similarly, there are instances in which the steam should be
considered as a by-product. The by-product steam method charges
the shaft work on the basis of the cost to produce shaft work from
a condensing turbine or on the cost of purchased shaft work (really
electricity plus the cost of a motor), whichever is cheaper:

$$c_{shaft} = c_{shaft,\ condensing}$$
or
$$c_{shaft} = c_{shaft,\ purchased}$$

The cost of low-pressure steam, c_{LP}, can then be obtained from the
turbine money balance. This scheme will yield a low unit cost of
steam and a high unit cost of shaft work.

The choice of a costing method depends upon the object of
economic study and/or upon the circumstances under which it is
being made.

Up to this point the methods discussed involve only average
unit costs of flow streams. A different problem is one in which,
for example, an economic study has the purpose of ascertaining the

desirability of increasing (or decreasing) the manufacture of some product, which results in additional power requirements. The incremental method--which charges the additional units of power required for the cost of producing them--is then appropriate. The use of the average cost of producing all of the units would be inappropriate (except in the special, though not infrequent, cases when incremental costs are close to average costs). Another instance when incremental costs are appropriate is in the case of design optimization.

Design Optimization. The goal in design optimization of energy conversion systems as considered here is to select the equipment which strikes the best balance between overall capital (and other) costs and the cost of the availability input for the particular type of system operating under the given constraints. That is, the aim is to minimize the total expenditure for capital and fuels. The rationale of the use of available energy in such optimization is that it allows a large complex system to be optimized by parts, splitting the overall system into much less complex subsystems (perhaps unit operations and/or individual components thereof) and optimizing these individually, rather than optimizing the whole system at once. In such situations, much insight can be applied because of dealing with the much simpler subsystems, leading to easier optimization.

The use of available energy in optimization of a complex energy system is quite similar to its use in costing of energy flows as presented above, inasmuch as the keys are (i) the use of money balances, and (ii) the assignment of unit costs to available energy. On this basis, a cost or value of the available energy flows at the various junctures within the system can be determined (iteratively). Then if, for example, a particular component needs to deliver some specified output, knowing the unit cost of the availability supplied to the component, the component and its operating conditions can be selected (or designed) so that the total costs--of availability supplied thereto and of capital costs thereof, (etc.)--can be minimized. Typical parameters (decision variables) of a component which can be adjusted in order to attain the optimum system are efficiency, operating pressure and temperature, speed, and so on and so forth.

Evans, Tribus and co-workers (14,15,16) have treated this subject under the title of Thermoeconomics. They express each explicit variable of the overall system money balance as a function of the design parameters--those variables which could be arbitrarily selected by the designer in order to obtain a workable design, but which ought to be rationally selected in order to obtain an optimal design. Applying the usual calculus methods, they determine the conditions which would minimize the total expenditure. It is shown that, as long as a system operates at steady (or quasi-steady) conditions, it is possible to separate a complex "energy" intensive system into subsystems which, when individually optimized by balancing costs of available energy dissipation against amortized capital costs, will optimize the overall system. As mentioned above, the key is that the proper value be assigned to the available energy at the various subsystem junctures. The work of Evans, et al., shows in general that these proper values are incremental costs, arising as adjoint Lagrange multipliers of the respective available energy streams. As expected, the values at junctures within the system are generally higher than the cost of available energy in the primary fuel inputs to the system.

Consider an example of Evans, et al., which consists of three subsystems in an overall system, as shown in Figure 3a (14). Figure 3b shows the system with the subsystems separated and ready for economic optimization with the λ's representing unit available energy values at the points indicated. To do this rigorously, the capital costs and input available energy rate of each subsystem must be expressed as functions of the output available energy rate and the decision variables $\{X\}$ particular to that subsystem.

Let us consider only two special cases among those considered by Evans, et al. (14). For the case in which the subsystem capital costs are insensitive to the output available energy rate and each subsystem cost depends only on the thermodynamic efficiency η_{II}, then the result of the mathematical optimization is that $\lambda_1 = c_1$, $\lambda_2 = c_1/\eta_{II,1}$, and $\lambda_3 = \dfrac{\lambda_2}{\eta_{II,2}} = \dfrac{c_1}{\eta_{II,1}\eta_{II,2}}$. These are the expected results since, if changing the output doesn't change the capital cost, the only incremental cost for increasing the output of a subsystem is the increase in "fuel" (i.e., available energy input) cost, and this latter cost is proportional to the unit cost of the "fuel" and inversely proportional to the efficiency of "fuel" utilization.

Consider the situation where the subsystem capital costs increase linearly with output available energy,

$$C_i = a_i + b_i P_{i+1}$$

and are functions of the variables $\eta_{II,1}$, $\eta_{II,2}$, and $\eta_{II,3}$ [i.e., $a_i = f_i(\eta_{II,i})$, $b_i = g_i(\eta_{II,i})$]. Then the costs are

$$\lambda_1 = c_1$$
$$\lambda_2 = b_1 + \frac{c_1}{\eta_{II,1}}$$

and

$$\lambda_3 = b_2 + \frac{c_2}{\eta_{II,2}} .$$

Notice that these costs also turn out as expected; the incremental cost is that associated with increased capital expenditure, proportional to output in the case at hand, plus that associated with

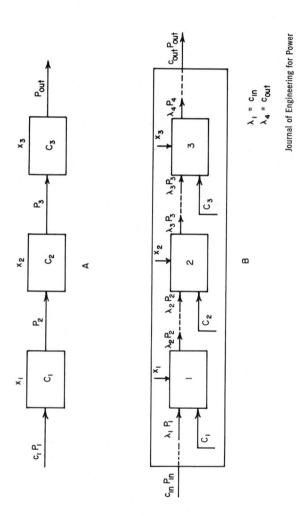

Journal of Engineering for Power

Figure 3.   System with three subsystems illustrating their separation (14)

increased "fuel" costs as before. Observe that, if $a_i = 0$ and b_i is constant, then $b_i = C_i/P_{i+1}$, and these costs are the same as the average costs used in the costing of energy utilities discussed. Many times the average costs adequately approximate the incremental costs, especially for devices in which the second-law efficiency is relatively constant. The restriction of the foregoing methods to steady systems can also be relaxed when the subsystem efficiencies change little with system load.

This second special case, with capital costs continuously proportional to subsystem output, is a reasonable approximation to the cost of equipment and installation for the usual subsystem. Therefore, it can be a quite useful assumption, especially for the first iterations of the design optimization. However, the usual subsystem cannot be purchased in continuously varying sizes but only in step-wise varying sizes with corresponding step-wise varying prices. Nevertheless, continuously varying subsystem outputs can be achieved by varying the operating parameters of a subsystem.) Therefore, the assumed proportionality can only be an approximation to the step-wise varying cost. The results of the optimization will then indicate which size of subsystem to install. Once the size is decided, then the optimization can be carried further, to determine the optimal operating conditions for this size of subsystem in the system at hand; this further optimization would follow the former special case, with capital costs independent of subsystem output.

It should be mentioned that other methods of design optimization, employing the Second Law for costing, have been used. For example, without explicitly determining the cost of available energy at each juncture of a system, in 1949 Benedict (see 19) employed the Second Law for optimal design. He determined the "work penalties" associated with the irreversibilities in an air separation plant. That is, the additional input of shaft power to the compressors required as a consequence of irreversibilities was determined from the entropy production in each subsystem. Associated with additional shaft power requirements are the costs of the power itself and the increased capital for larger compressors. Each subsystem could then be optimally selected on the basis of the tradeoff between increased capital expenditure for reducing subsystem irreversibilities on the one hand, and increased power and compressor costs on the other hand.

Obert and Gaggioli (6) applied Second Law costing to the optimal design of steam piping and its insulation in a power plant. In essence, they costed the consumption and loss of the steam's available energy in terms of the lost output of power. More recently Wepfer, Gaggioli and Obert (13) presented the methodology for economic design of piping and insulation on the basis of available energy costs determined by the methods described earlier in the present article.

The application of Second Law methods to design is in its infancy. However, it is clear from the few simple cases where it has been used for these purposes, and for decision-making in system maintenance and operation, that further development of available energy accounting for optimization is important. This development can best be achieved by tackling additional practical applications, beginning with the more manageable ones and proceeding to the more difficult ones as individual and collective experience is gained.

   <u>Maintenance and Operation Decisions</u>. The determination of an appropriate cost of available energy at various junctures of a system in a manner similar to that described above for <u>Design Optimization</u> is useful not only in design but also allows <u>decisions</u> regarding the repair or replacement of a specific subsystem to be readily made (<u>9,10</u>). The amortized cost of such improvements can be easily compared with the cost of the additional available energy that will be dissipated if a component is left to operate in the given condition. The proper decision then becomes very apparent.

   The cost of available energy that should be used in these maintenance applications is determined in a manner very similar to those described in the foregoing section. However, the exact procedure of treating capital costs depends on the nature of the overall system. If the system is to produce at a constant output and is presently able to do that, then the deteriorated condition of a specific component is merely resulting in extra usage of fuel available energy. In this situation, the sunk capital costs do not influence the unit availability costs at each juncture, and capital costs should be disregarded. For a series system, the unit availability costs would then start at the base fuel-available-energy cost and merely be increased by a factor equal to $1/n_{II,i}$ at each component i. If, on the other hand, an increased market is available for output from the system, then the unit available energy prices should be set by beginning with the output availability, setting its (incremental) unit cost equal to the market price. Then, proceeding backwards through the system, the unit cost of the input to subsystem i is $n_{II,i}$ times that of t'e output therefrom.

Conclusions

   It has not been the intent of this paper to give a thorough treatment of the theory of available energy accounting. Rather, the aim has been to introduce the subject simply, presenting the relationships and methods for applying such costing in simple applications and describing the various practical uses of Second-Law costing. If further elaboration on the fundamentals is desired, the reader may refer to (<u>12,14</u>) for examples.

   Neither has it been our purpose to present the specifics of one (or more) sample applications. For examples of applications in detail, see (<u>9,10,12,13,19</u>) and the papers by Wepfer, Benedict and Gyftopoulos, and Petit in this volume.

   Thus, it is hoped that this paper and the companion paper by Wepfer will encourage and facilitate the practical use of available-energy accounting by enhancing the understanding of the basic concepts.

In summary, available-energy costing can greatly facilitate (i) cost accounting of "energy" commodities, and (ii) the economics of alternative choices in design and operation of a system and/or its components. Often the same results can be achieved without recourse to available energy, but only through more difficult, indirect procedures.

Nomenclature

Symbol

A_c	-	availability destruction (irreversibility)
A_Q	-	availability of heat transfer (positive into the system)
A_W	-	availability of work transfer (positive into the system)
a_f	-	availability per unit flowing mass (flow availability)
C	-	annual cost rate ($/year for example)
c	-	unit cost
E_Q	-	energy of heat transfer
E_W	-	energy of work transfer
e_f	-	energy per unit flowing mass
h	-	enthalpy per unit mass
m_f	-	mass flow
P^f	-	power, typically kWh/year
p	-	pressure
p_0	-	dead state pressure
s	-	entropy per unit mass
T	-	temperature
T_0	-	dead state temperature
V	-	volume of system
W	-	work transfer
X	-	decision variables of a subsystem
λ	-	unit cost of available energy as Lagrange multiplier
$\mu_{j,o}$	-	dead state chemical potential of constituent j
η_I	-	first law efficiency
η_{II}	-	second law efficiency

Subscript

e	-	exit
i	-	inlet
0	-	dead state condition

Overline

$\cdot$	-	per unit time

Literature Cited

1. **Maxwell, J.** C.,"Theory of Heat," Longmans Green, London, 1871
    (also see later editions).
2. Gibbs, J. W., 1875; see "Collected Works," vol. 1, p. 77,
    Yale U. Press, 1948.
3. Keenan, J., "A Steam Chart for Second Law Analysis," Trans.
    A.S.M.E., 1932, 54, (195).
4. Keenan, J. H., "Thermodynamics," John Wiley and Sons, Inc.,
    New York, 1941.
5. Obert, E. F., "Thermodynamics," McGraw-Hill, New York, 1948.
6. Obert, E. F. and Gaggioli, R. A., "Thermodynamics," McGraw-Hill,
    New York, 1963 (Second Edition).
7. Gaggioli, R. A., "The Concept of Available Energy," Chem. Engr.
    Sci., 1961, 16, (87).
8, Reistad, G. M., "Availability: Concepts and Applications,"
    PhD Thesis, University of Wisconsin, 1970.
9. Fehring, T. and Gaggioli, R., "Economics of Feedwater Heater
    Replacement," Trans. A.S.M.E., J. Eng. Power, 1977, 99.
10. Gaggioli, R. and Fehring, T., "Economics of Boiler Feed Pump
    Drive Alternatives," Combustion, 1978, 49, (9). Also ASME
    Paper No. 77/JPGC-PWR-8.
11. Gaggioli, R. A., "Proper Evaluation and Pricing of Energy,"
    Proc. Int. Conf. on Energy Use Management, 1977, II,
    Pergamon Press.
12. Gaggioli, R. A., and Wepfer, W. J., "Available Energy Account-
    ing - A Cogeneration Case Study." Presented at the A.I.Ch.E.
    Annual Meeting, June 1978.
13. Gaggioli, R., and Wepfer, W., "Economic Sizing of Piping and
    Insulation," to be presented at ASME Winter Annual Meeting,
    San Francisco, Dec. 11-14, 1978.
14. Evans, R. and El-Sayed, Y., "Thermoeconomics and the Design
    of Heat Systems," Trans. A.S.M.E., J. Eng. Power, 1970, 92,
    27-35.
15. Tribus, M., Evans, R., and Crellin, G., "Thermoeconomics,"
    Ch. 3 of "Principles of Desalinatio," edited by K. Spiegler,
    Academic Press, 1966.
16. Tribus, M., and Evans, R., "Thermo-economics," UCLA Report
    No. 52-63, 1962.
17. Petit, P., and Gaggioli, R., "Second Law Procedures for Evalu-
    ating Processes," this volume.
18. Rodriguez (S.J.), L., and Petit, P., "Calculation of Available-
    Energy Quantities," this volume.
19. Gyftopoulos, E., Widmer, T., and Lazaridis, L., "Potential Fuel
    Effectiveness in Industry," Bollinger Publishing Co., Cambridge,
    MA, 1974.
20. Wepfer, W., "Applications of Availability Accounting," this
    volume.

RECEIVED October 30, 1979.

Applications of Available-Energy Accounting

WILLIAM J. WEPFER[1]

Professional Engineering Consultants, 3915 N. Lake Drive, Milwaukee, WI 53211

This paper surveys the results of three practical applications of available-energy costing as well as some of the accom-
panying methodology. To illustrate the use of available-energy
accounting in the costing of system products, the results of a
study made for a firm (a paper manufacturer) co-generating shaft
power (and electricity) as well as low-pressure process steam for
different end-uses is shown. Comparison with the firm's previous
costing of excess electricity emphasizes the crucial importance
of second-law methods. It should be mentioned that these same
techniques can be applied to any system with multiple products,
for examples: (i) co-generation systems having hot-water, compressed air or refrigeration outputs in addition to steam and
shaft power; and, (ii) process industries like petroleum refining
engaged in the production of many products.

The application of available-energy costing to facilities
operation is illustrated with an example from the electric utility
industry--namely that of feedwater heater maintenance and replacement. Here available-energy costing holds the key to determining when the heater should be replaced altogether, after
successively plugging leaky tubes.

The final example applies available-energy costing to an
optimal design problem--the economic sizing of steam piping and
insulation. It is only through the available-energy concept that
the relative dissipations due to pipe friction and heat transfer
can accurately be assessed. In turn, the true costs of these
dissipations can then be weighted against the capital costs of
piping and insulation.

Results of Costing Co-generated Steam and Shaft Power

Co-generation is a vital element in the paper industry because paper production requires large quantities of low-pressure

[1]Current address: Professional Engineering Consultants,
Milwaukee, WI 53211.

0–8412–0541–8/80/47–122–161$06.50/0
© 1980 American Chemical Society

steam and mechanical shaft power. The following results, taken
from a case study in the paper industry (1), were obtained by
applying the available-energy costing methodology described in
the preceding paper by Reistad and Gaggioli (2) to three representative co-generating power plants. These results are contained
in Table I and on Figures 1-3.

Figure 1 is a simplified schematic diagram of a co-generation
system typical of systems built twenty years ago. (In fact, this
system will be designated as the 1960 system.) The diagram shows
the average annual available-energy flows and consumptions as well
as the unit costs of process steam and electricity outputs.

The incentive for improving energy conversion systems is generally economic. The capital cost to improve a system must result
in a savings which is greater. The 1960 system was built in a
time when fuel was inexpensive and there was little or no incentive to build efficient systems. The overall second-law efficiency of 28% is indicative of this fact.

The so-called 1976 system is shown in Figure 2. This system
utilizes a coal-fired boiler to generate 600 psi steam. The increase in overall efficiency to 33% is primarily due to the elimination of the large heat transfer consumptions associated with the
low-pressure boiler in the 1960 system.

The paper company, contemplating the addition of a paper
machine to increase their production, sought to expand their
co-generating capacity. The 1980 system, as illustrated in Figure 3, has been proposed as the typical means of supplying the
required steam and shaft work. (See reference (3) for other more
economic alternatives.)

In the 1980 system, steam would be produced at 1250 psi.
Under these conditions, the temperature difference between the
combustion products and the steam is significantly reduced, and
the consumption of available energy that drives the heat transfer
is decreased to 23% of the fuel input. This bodes well for the
efficiency of the system as a whole. The overall efficiency of
35% for the 1980 system is really quite good when compared to most
energy systems and represents a significant improvement over the
28% efficiency of the 1960 system.

The products of each system were costed using two different
techniques: the equality method and the by-product work method
(1,2). Every case analyzed accounts for maintenance as well as
fuel expenses but excludes other operating costs. Capital investment is amortized at an after-tax discount rate of 8.5%. The effects of income taxes, ad valorem taxes, and depreciation (see
(3) for the formulation of the annualization factor accounting for
capital, taxes and depreciation) are included, while the effects
of inflation were neglected.

The 1960 system was analyzed without accounting for amortization on the assumption of sunk capital, inasmuch as the system
has been almost fully depreciated. In addition to analyses including amortization, costings for the 1976 and 1980 systems were

Table I. Unit Available Costs for Cogeneration Systems

System	Capitalization	Cost Distribution Method	65 Psig Steam Cost $/10^6 Btu	65 Psig Steam Cost ¢/kWh	175 Psig Steam Cost $/10^6 Btu	175 Psig Steam Cost ¢/kWh	Average Steam Cost $/10^6 Btu	Average Steam Cost ¢/kWh	Electricity ¢/kWh Back-pressure	Electricity ¢/kWh Condensing	Electricity ¢/kWh Average
1960	Sunk	Equality	5.22	1.78	5.22	1.78	5.22	1.78	2.01	---	2.01
		By-product	6.12	2.09	5.97	2.04	6.07	2.07	1.08	---	1.08
1976	$9.1(10^6) 8.5% 20 yrs	Equality	7.86	2.68	7.86	2.68	7.86	2.68	3.02	---	3.02
		By-product	10.0	3.42	7.79	2.66	9.24	3.15	1.74	---	1.74
1976	Sunk	Equality	5.29	1.81	5.29	1.81	5.29	1.81	2.04	---	2.04
		By-product	7.02	2.40	5.25	1.80	6.39	2.18	0.97	---	0.97
1980	$20.5(10^6) 8.5% 20 yrs	Equality	6.14	2.1	6.14	2.1	6.14	2.1	2.40	4.19	2.64
		By-product	8.44	2.88	6.5	2.23	8.17	2.79	1.52	4.43	1.92
1980	$20.5(10^6) 8.5% 32 yrs	Equality	6.08	2.07	6.08	2.07	6.08	2.07	2.37	4.13	2.61
		By-product	8.37	2.86	6.47	2.21	8.10	2.76	1.49	4.36	1.88
1980	Sunk	Equality	4.80	1.60	4.80	1.60	4.80	1.60	1.78	2.86	1.93
		By-product	7.04	2.40	5.27	1.81	6.79	2.30	0.92	3.12	1.22

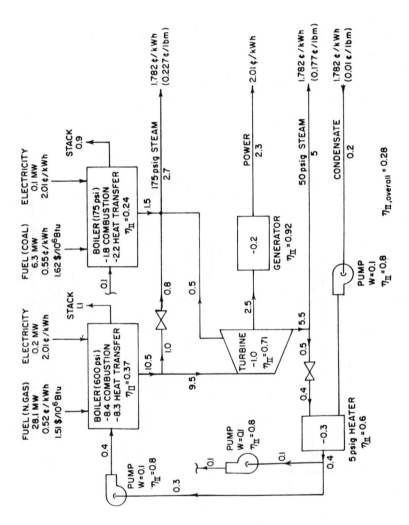

Figure 1. Skeleton schematic of 1960 system available-energy flows (megawatts).
Also shown are the unit costs of the steam and electricity outputs based on the
equality method and sunk capital costs.

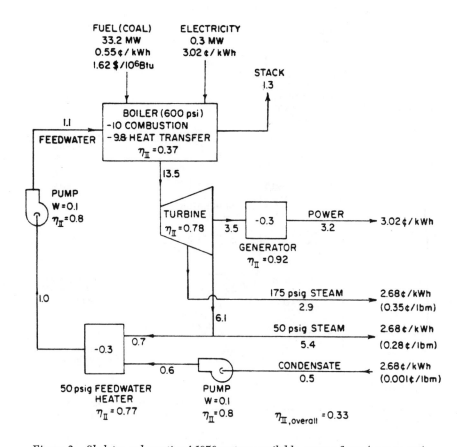

Figure 2. Skeleton schematic of 1976 system available-energy flows (megawatts). The unit costs of steam and electricity were computed using the equality method and a capital charge of $9.0(10^6) at an effective after-tax interest rate of 8.5% and an economic life of 20 years.

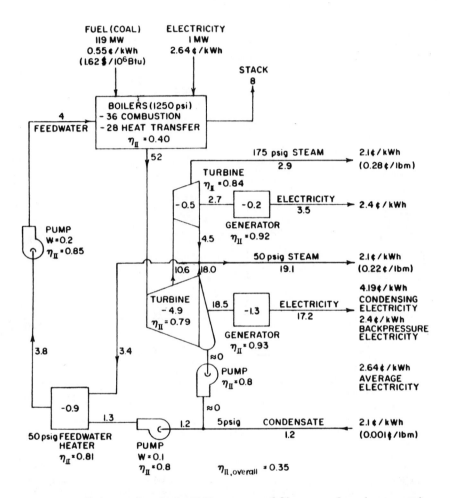

Figure 3.   Skeleton schematic of 1980 system available-energy flows (megawatts). The unit costs of steam and electricity were calculated with the equality method and for a capital charge of $20.5(10⁶) at an effective after-tax interest rate of 8.5% and an economic life of 20 years.

also done assuming sunk capital since these results are useful
for some economic considerations.

What criteria should be used when deciding which costing
method is appropriate? If a company were in the position where
it "needed" to produce its own steam (because it could produce it
more cheaply, in boilers, than it could purchase it), then the
by-product work method would be appropriate for establishing the
desirability of investing in a co-generation system. If the resultant average cost of by-product electricity is less than it
could be purchased for, the investment in co-generation facilities
would be worthwhile. This is one example where the by-product
work method would be appropriate, for making <u>internal</u> decisions.

It should be mentioned that the conventional technique for
costing process steam and shaft work outputs of a co-generating
facility is nearly equivalent to the by-product work method (4).
In the conventional method the shaft power is charged with the
incremental fuel cost associated with producing it as well as the
capital charge for the turbine. (This is closely equivalent to
charging steam as if it were produced in a low-pressure boiler.)
The cost of process steam is then obtained from a money balance
on the turbine. Nevertheless, the commodity of value is available energy--any method which assigns costs on any other basis
such as energy or mass is <u>usually invalid</u>. Furthermore, only with
available-energy costing can co-generating power plants be analyzed by other methods--equality, extraction, by-product steam--
discussed in the preceding paper (2).

Once a company has decided to produce steam and shaft work
from back-pressure turbines, say because of overall economics,
then is it still reasonable to use the by-product work method?
The by-product work method makes electricity appear to be relatively cheap and steam relatively expensive, compared to the equality
method. After fixed expenses for the system are sunk, then the
costs associated with delivering available energy in back-pressure
steam and in turbine shaft power are the same. (This contention
is fortified by the fact that the marginal cost for useful energy
in such steam and in shaft output are the same, equal to the average costs, insofar as the efficiencies with which power plant
components operate remain practically constant (5). That is, the
cost of delivering another unit of available energy in steam
equals the average cost of the units already being delivered in
the steam output and shaft output. And, likewise for an additional unit of shaft power.)

Suppose that a paper mill were purchasing all the electricity
<u>and</u> steam it used, say because the steam and electric generating
equipment all belonged to another concern--a utility. The
utility would--or at least should--charge for the steam and electricity in proportion to the cost of producing each. Since for
the utility neither is a by-product of the other, the cost of
producing the back-pressure steam and the turbine shaft power is
proportional to their available-energy content; this calls for
the equality approach.

168 THERMODYNAMICS: SECOND LAW ANALYSIS

For the case study at hand, the company was engaged in the sale of "excess" electricity to outside customers. Using energy-based costing this company, even after the allowed mark-up for a profit, was selling electricity below cost--at 1.8¢/kWh, even lower that the 2.01¢/kWh cost for producing the power with the 1960 system assuming <u>sunk</u> capital--while overcharging the steam going to its own processes. Even before this study was commissioned, the company realized it was underselling electricity but could not argue its case rationally to the utility commission. Clearly this is a situation in which energy costing causes errors which are avoided when available energy is taken as the commodity of value.

Note, too, that while the company's decision to co-generate power (along with the steam that it needed to produce) might have been based on a <u>by-product work</u> analysis; once engaged in the production of electricity and its sale to outside consumers, the company should (and would prefer to) use the equality method (or, better, the extraction method) for costing power--at least the power to be sold.

Another important conclusion was drawn from the results of available-energy costing. The results for the 1980 system showed that the cost of the power produced by the "condensing stage" of a condensing turbine was much more than that of purchased power (∿ 2.35¢/kWh at the time of this study). It can be said that new condensing power should not be installed by the company--unless it can be justified from the standpoint of "peaking economics" (or some other necessity). If condensing turbines are justified by peaking needs, then the "condensing stage" should be a separate turbine which can be shut down when not needed for peaking since "condensing power" is more expensive than purchased power, even when capital is sunk.

The Economics of Feedwater Heater Replacement

Second-law techniques are applied to the problem of determining when to replace feedwater heaters. This analysis was performed by Fehring and Gaggioli (6) as a test of a decision which had recently been made by a utility. The details of the analysis, omitted here, are contained in reference (6).

Modern steam power plants employ feedwater heaters to raise the temperature of the feedwater before entering the boiler. Steam is extracted from the turbines at various "bleed points" and is condensed over tube bundles inside which the boiler feedwater runs; Figure 4.

Feedwater heaters increase plant efficiency by increasing the average temperature at which heat is supplied to the cycle H_2O from the products of combustion (7, articles 15.1-6); that is, inasmuch as the available-energy consumption for driving heat transfer processes increases with the temperature difference

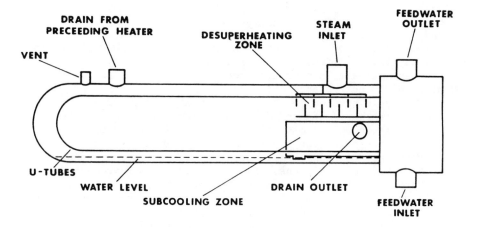

Figure 4.   Typical feedwater heater

(really the temperature gradient), feedwater heaters increase plant efficiency by decreasing the average ΔT for heat transfer to H_2O (5).

The number of feedwater heaters in the cycle is commonly based upon the economic trade-off between the first cost of the unit and the expected annual operating savings due to the increase in unit efficiency. In practice, it is also limited by the maximum number of bleeder stages which the turbine manufacturer supplies with a given size turbine.

As feedwater heaters age, there is film build-up inside and outside the tubes, and leaks invariably occur. Normally, leaks occur infrequently during the first decade of the heater's life with the frequency increasing rapidly as the mean life of the tubes is approached. A heater is "repaired" simply by plugging the leaky tubes.

If the heater is conservatively designed, as is usually the case, the initially plugged tubes cause small, if any, deterioration of performance. Eventually however, film buildup and plugged tubes increase to the point that the heater can no longer operate near design conditions. As the feedwater heater efficiency drops off, it affects the overall cycle efficiency, which causes an increase in operating costs; and, at some stage, the heater should be replaced.

The cycle studied in the ensuing analysis is a fairly typical one, with seven stages of extraction to the feedwater heaters. This unit is shown schematically in Figure 5. The properties of H_2O at various points in the feedwater system as well as the mass flow rates are given in reference (6). Three cases are analyzed, Case A at design conditions, while Cases B and C represent deteriorated conditions to be described presently.

A complete analysis of useful energy flows, consumptions and losses for this unit under design conditions, Case A, has been presented earlier (5), and the results are employed in the economic analysis to follow.

Case B is intended to represent the situation when, after years of service, the heater supplied by B5 has had, say, twenty percent of its tubes plugged. The terminal temperature difference (TTD--the difference between saturated steam temperature and feedwater outlet temperature) can be expected to increase by approximately five degrees (^{O}F). If heater number 6, supplied by B6 is in good condition it would nearly pick up the load which heater 5 failed to carry (but less efficiently). These are the circumstances assumed for Case B.

At some stage the deterioration of performance becomes so great that the heater should be taken out of service in order to either (i) replace it with a new heater, or (ii) re-tube the heater, which generally involves more down-time than replacement. Case C portrays the situation when heater 5 is out of service, causing greater upset to the feedwater circuit than Case B. Notably, the temperature of the feedwater to the boiler is lower,

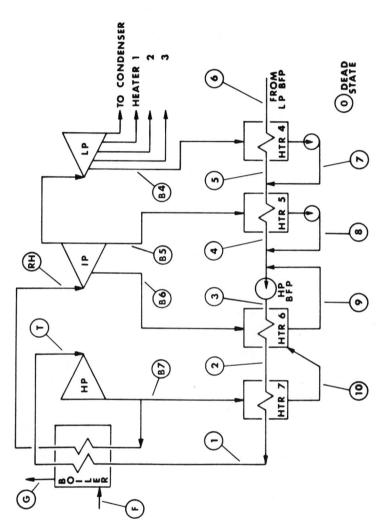

Figure 5.   Schematic of typical unit

and additional boiler fuel must be burned in order to maintain the required system output.

The economic analysis to follow depends upon the evaluation of the various available-energy supplies for feedwater heating and, in turn, the costs associated with those supplies. In particular, the costs of interest, for each case, are those required to take the feedwater from the conditions at the inlet to heater number 4 to the normal temperature of feedwater entering the boiler. These costs include the cost of bleeder steam, which is used to increase the temperature of feedwater in the heater; and, under the conditions of Case C, the cost of the additional boiler fuel required when the heater is out of service and the temperature of the feedwater is below normal. The hourly cost of feedwater heating for Cases A and B is given by

$$\$/hr = \sum_{i=4}^{7} c_{Bi}\dot{A}_{Bi}$$

while that for Case C is

$$\$/hr = \sum_{i=4}^{7} c_{Bi}\dot{A}_{Bi} + c_F \dot{A}_{F,add}$$

where $\dot{A}_{Bi}$ is the available-energy flow in the ith bleed line, $\dot{A}_{F,add}$ is the additional boiler fuel required when the heater is out of service, c_{Bi} is the unit cost of steam in the ith bleed line, and c_F is the unit cost of boiler fuel (coal). These available-energy flows are calculated in the usual manner and are given in Table II and reference (6). The unit costs are obtained with exactly the same techniques discussed by Reistad and Gaggioli (2)--by applying money balances to each component (boiler, high-pressure turbine, . . .) of the system of interest to get the dollar flows and then by the subsequent division of the appropriate available-energy flows to obtain the unit costs.

$$c_{product} = c_{supply}[\dot{A}_{supply}/\dot{A}_{product}] + \dot{\$}_{capital}/\dot{A}_{product}$$

There are two components to the cost of product available energy; one contribution from fuel expense

$$c_{supply}[\dot{A}_{supply}/\dot{A}_{product}]$$

and one from capital investment

$$\dot{\$}_{capital}/\dot{A}_{product}.$$

The available-energy costs employed in the present analysis do not include capital expenses. Capital costs are not neglected; they

are irrelevant because the capital for all the equipment involved
in supplying available energy to the feedwater heaters is already
sunk. Table II lists the unit costs for each steam bleed line
supplying available energy to the feedwater heaters.

Economic Justifiability of Heater Replacement. On the basis
of the hourly costs of feedwater heating shown in Table II, the
difference in operating (fuel) costs between the deteriorated
heater and that of a heater functioning at design conditions (Case
A) can be ascertained and weighted against the investment required
for replacing or retubing the heater. It is assumed that the deteriorated heater operates under the circumstances of Case B but
with an annual down-time to allow for the plugging of tubes (Case
C).
If the unit operates 8000 hours per year at a 70% capacity
factor, the calculated additional annual fuel cost due to the deterioration of the heater No. 5 is:

$$\begin{bmatrix} \text{Additional fuel} \\ \text{cost during op-} \\ \text{eration} \end{bmatrix} = [\text{hr/yr}] \begin{bmatrix} \text{capacity} \\ \text{factor} \end{bmatrix} \begin{bmatrix} \text{Case B} & \text{Case A} \\ \text{hourly} & - & \text{hourly} \\ \text{cost} & \text{cost} \end{bmatrix}$$

$$\$10438/yr = [8000 \text{ hr/yr}][0.70][\$295.957/hr - \$294.093/hr]$$

(The use of a simple capacity factor is tantamount to assuming
that the hourly costs vary in proportion to the load on the unit.
This is a good assumption, since the second-law efficiencies,
underlying the unit costs, do not vary much with load (5).) And,
if the heater is out of service for 3 weeks per year for the
plugging of heater tubes, the annual heater down-time fuel cost is:

$$\begin{bmatrix} \text{Additional fuel} \\ \text{cost during} \\ \text{down-time} \end{bmatrix} = [\text{down-time}] \begin{bmatrix} \text{Case C} & \text{Case A} \\ \text{hourly} & - & \text{hourly} \\ \text{cost} & \text{cost} \end{bmatrix}$$

$$\$12,688/yr = [3 \text{ wk/yr}][168 \text{ hr/wk}][\$319.267/hr - \$294.093/hr]$$

In addition to the additional fuel costs accrued during heater
outage, a manpower expenditure is required to plug the leaks. If
15 leaks per year occur and if 28 man-hours are charged at
$10.07/man-hour during each heater outage, the annual additional
maintenance expenditure is:

$$\begin{bmatrix} \text{Additional} \\ \text{maintenance} \\ \text{expenditure} \end{bmatrix} = [\text{outages/hr}][\text{man-hr/outage}][\$/\text{man-hr}]$$

$$\$4229/yr = [15 \text{ outages/yr}][28 \text{ man-hr/outage}][\$10.07/\text{man-hr}]$$

Then the total annual additional fuel and maintenance expenditure

Table II. Available Energy Requirements and Costs

Useful Energy Source	Useful Energy Requirements			Unit Cost of Useful Energy $/10^6 BTU	Cost of Useful Energy Requirements		
	BTU/HR (J/s) CASE A	BTU/HR (J/s) CASE B	BTU/HR (J/s) CASE C		$/HR CASE A	$/HR CASE B	$/HR CASE C
B4	37,818,000 (11,083,000)	37,818,000 (11,083,000)	39,069,000 (11,450,000)	1.270	48.028	48.028	49.618
B5	27,787,000 (8,144,000)	24,824,000 (7,275,000)	-	1.354	37.624	33.611	-
B6	49,337,000 (14,459,000)	53,365,000 (15,640,000)	72,080,000 (21,124,000)	1.459	71.983	77.860	105.165
B7	94,565,000 (27,714,000)	94,565,000 (27,714,000)	99,050,000 (29,028,000)	1.443	136.457	136.457	142.929
F Increase	-	-	26,945,000 (7,897,000)	0.800	-	-	21.556
Totals	209,507,000 (61,400,000)	210,573,000 (61,712,000)	231,620,000 (67,880,000)		294.093	295.957	319.267

due to heater No. 5 is:

$27,355/yr = $10,438/yr + $12,688/yr + $4,229/yr

Any number of methods for economic analysis could be employed;
here, a discounted cash flow analysis is used. Assume that: (a)
the heater replacement cost is $235,000; (b) the replacement heat-
er will have a 20-year life and an $18,000 salvage value; (c) fuel
and maintenance savings are escalated at 6% per year; (d) after
tax cost of capital is 9%. Table III shows the results of the
cash flow analysis of the proposed investment in a replacement
heater. As is indicated, since the net cumulative discounted
cash flow (+ $29,166) is positive, the replacement of the heater
is justified.

Alternatively, the heater might be retubed rather than replaced. However, it then may have to be removed from service for
an extended period of time. If the heater could be retubed for
$185,000 (vs. $235,000 for replacement), a savings can be realized
if retubing can be accomplished in less than:

[$235,000-$185,000]/[319.267-294.093]$/hr = 1986 hr = 11.8 weeks

where the denominator is the additional fuel expenditure due to
heater down-time.

Because this is an operational system, capital charges for
existing equipment are irrelevant. However, it should be noted
that other differential costs may be applicable to the deterioration of heater performance. For example, if the cycle efficiency
decreases greatly, and if the boiler does not have sufficient
overcapacity, the loss of unit efficiency will cause a decrease in
maximum output and, hence, could require the expenditure of additional capital. This would result in an added monetary benefit
possible through replacement of the feedwater heaters in question.

While the analysis presented here was for a plant operating
problem, the same methods could be applied at the design stage
for ascertaining the desirability of prospective improvements--
for design optimization. The principles are not new--Benedict
(8) used them in 1949 to optimize the design of an oxygen separation plant. In the next section, the application of these techniques to the selection of steam piping and insulation will be
discussed.

Economic Sizing of Steam Piping and Insulation

The determination of the economic choice for pipe size and
insulation thickness has been discussed by many authors (9,10,11
12). Most piping systems are designed on the basis of simple
rules-of-thumb in which the primary factor is a velocity or a
pressure gradient. Often times the pipe is sized prior to any
consideration of insulation thickness, which is in turn selected

Table III. Cash Flow Analysis, Feed Water Heater Replacement Analysis

Year	0	1976	1977	. . . .	1995
Fuel and Maintenance Saving		$ 27,355	$ 28,996		$82,764
- Depreciation (S.Y.D.) of Replacement Heater		20,000	19,111		3,111
- Ad Valorem Taxes		5,640	5,499		2,961
= Taxable Balance		$ 1,715	$ 4,386		$76,692
- Income Taxes		858	2,193		38,346
+ Investment Tax Credit	$ 23,500				
= After Tax Balance	$ 23,500	$ 858	$ 2,193		$38,346
+ Depreciation		20,000	19,111		3,111
+ Salvage Value of Replacement Heater					18,000
- Heater Replacement Cost	235,000				
= Total Cash Flow	-$211,500	$ 20,858	$ 21,304		$59,457
x Present Worth Factor (@ 9%)	1.0	.9174	.8417		.1784
= Discounted Cash Flow	-$211,500	$ 19,134	$ 17,932		$10,607
Cumulative Total of Discounted Cash Flow	-$211,500	-$192,366	-$174,434		+$29,166

to minimize the overall cost of insulation and "lost heat energy."
The underlying principle of these methods--based upon the First Law
of Thermodynamics--is the assignment of an economic value (cost)
to energy losses associated with heat transfer and pipe friction.
The result of such analysis generally is <u>not</u> an optimal design.
The purpose of this discussion is to review a method which guaran-
tees the optimal selection of pipe size and insulation thickness.
    The problem to be treated here is the optimal sizing of a
steam line and its insulation as discussed by Obert and Gaggioli
(7,13). As a specific illustration, a bleeder line which delivers
steam from a turbine to a feedwater heater will be designed. The
flow process is accomplished at the expense of available energy in
steam from the turbine; some of that available energy is consumed
to drive steam through the pipe. Furthermore, some is lost by heat
transfer through the pipe and its insulation. The optimal combination of pipe size and insulation thickness would be that which
minimizes the overall costs, which consist of: (1) the cost of
the available energy consumed and lost, plus (2) the capital (and
other) expenses for the piping and insulation. That is, the economic trade-off is the classical one between (1) "fuel" and (2)
"investment." To evaluate the "investment" costs requires the
determination of capital, maintenance, etc., costs for piping and
insulation. Needed to calculate the "fuel" (i.e., available energy) expenses are the amount of available energy consumed and lost,
as well as the unit cost of available energy.
    Consider Figure 5 again. Bleeder steam line B5 extracts
55351 lbm/hr (6.974 kg/s) at 93.8 psia (0.65 MPa) and 603 F
(317 C) from the intermediate-pressure turbine delivering it to a
feedwater heater. Steam line B5 was chosen to illustrate the design procedure for two reasons. First, the design has sufficient
flexibility; there are no extraordinary constraints on the pipe
material and size, or on insulation thickness. Secondly, the
available energy analysis and unit-cost computations for this
power plant, and for steam line B5, have been presented in previous papers (5,6).
    The function to be optimized, called the objective function,
is the yearly total cost of the piping system--the sum of the
amortized capital cost of the pipe and insulation plus the operating costs

$$\phi_T(D_n, \theta) = \phi_P(D_n) + \phi_I(D_n, \theta) + \phi_F(D_n) + \phi_Q(D_n, \theta)$$

where ϕ_P is the amortized pipe cost, ϕ_I is the amortized insulation cost, ϕ_F is the operating cost resulting from viscous dissipation, and ϕ_Q is the operating cost from heat losses. The
greater the investment in the pipe ϕ_P, the smaller the operating
cost from friction ϕ_F, the greater the expenditure for insulation
ϕ_I, the smaller the operating costs of heat losses ϕ_Q. The goal of
the optimization is to obtain an investment that minimizes the

annual total cost, ϕ_T.

The amortized pipe cost ϕ_P, is equal to the product of the capital cost of the pipe, Z_P, and an amortization factor, ψ (3);

$$\phi_P(D_n) = \psi \times Z_P(D_n)$$

Figure 6 shows the layout of the piping system. Equipment and installation costs (including welding) were obtained from manufacturer's data and from estimating guides for several nominal pipe diameters. These estimates were then fitted to obtain a formula for the installed pipe cost as a function of nominal pipe diameter, $Z_P(D_n)$.

The amortized insulation cost, ϕ_I, is equal to the product of the installed capital cost of the insulation, Z_I, and the amortization factor, ψ:

$$\phi_I(D_n,\theta) = \psi \times Z_I(D_n,\theta)$$

The cost of calcium silicate insulation is a function of the nominal diameter of the pipe and the insulation thickness. Costs for several pipe diameters and thicknesses (in half-inch increments) were obtained from quotations from a reputable insulation vendor. These estimates established the functional dependence, $Z_I(D_n,\theta)$, of insulation cost on thickness and pipe diameter.

The annual cost associated with friction is given by

$$\phi_F(D_n) = c_{st}\dot{A}_f$$

where $\dot{A}_f$ is the rate of destruction of available energy as a result of friction, and c_{st} is the unit cost of the available energy in the steam. It is the available-energy unit cost that is crucial to this technique. Reference (6) evaluates these unit costs for Line B5. For the case in which the power plant fuel (coal) costs \$0.80/10^6 Btu (\$0.948/10^6kJ), the value of the available energy in the steam at Bleeder No. 5 is \$1.354 per million Btu of available energy (\$1.284/10^6 kJ).

The available-energy destruction due to friction is computed from the relation (7)

$$\dot{A}_f = \frac{T_0}{T_{st}} \frac{\dot{m}V^2}{2g_c} \frac{L}{D_P} [f + \frac{1}{L}\Sigma_i K_i]$$

where f is the friction factor, which depends upon the pipe roughness and on the Reynolds number, Re. Given the steam conditions, the pipe roughness, and flow rate, the Reynolds number depends only on pipe diameter D_P. The length of pipe is L, while K_i is the head-loss coefficient for the ith fitting or valve. The flow in the pipe will be turbulent, and the friction factor can be evaluated from a Moody diagram or an empirical relation such as the

Colebrook equation. The head-loss coefficients are obtained from tabulated engineering data.

The annual cost of available energy lost by heat transfer through the pipe and insulation is

$$\phi_Q(D_n,\theta) = c_{st}\dot{A}_Q$$

where $\dot{A}_Q$ is the steady rate of available energy lost from the steam. The available energy loss is equal to the heat loss times the factor $[1 - (T_0/T_{st})]$;

$$\dot{A}_Q = [1 - T_0/T_{st}]\dot{Q}$$

where $\dot{Q}$ is the heat conducted through the insulation. Reference (13) contains the equations needed to solve for Q.

Given the foregoing expressions for evaluating ϕ_P, ϕ_I, ϕ_F, and ϕ_Q as function of D_n and θ, the pipe size and insulation thickness that minimize the annual total cost can be found. Reference (13) discusses techniques to solve for the optimal D_n and θ.

The Optimal Insulation Thickness and Pipe Diameter. The principal results are embodied in Figures 7 and 8. Figure 7 shows total cost curves as functions of nominal pipe diameter for three insulation thicknesses (θ = 1.0, 3.5, and 7.0 inches). Figure 8 illustrates the piping system costs (friction, heat-loss, pipe, insulation, and total) as functions of nominal pipe diameter when the insulation thickness is optimal for that pipe diameter.

The optimal combination for this design problem is a nominal pipe diameter of 12 inches, and an insulation thickness of 3.5 inches. The minimum annual total cost corresponding to this selection is $67.13/ft-yr.

The relative magnitudes of the various costs are illustrated in Figure 8. For this particular design problem it is apparent that the piping and friction costs are quite large (e.g., 94.4% = [$14.73/ft-yr + $48.64/ft-yr]/[$67.13/ft-yr] of the total cost for the optimal piping system) and that the heat-loss and insulation costs are of secondary importance. The piping cost is by far the largest and represents 72.4% ($48.64/$67.13) of the total cost for the optimal piping system. The reason for the large pipe cost is evident by inspection of Figure 6. The bleeder steam line requires many expensive valves, joints, and fittings; for instance, the expansion joint represents nearly 52% of the total pipe cost for the 16-inch system. To keep this problem in perspective, the total installed cost of $620.26/ft for the 16-inch system should be contrasted with a total installed cost of $35.00/ft for an equivalent length of straight 16-inch pipe. The valves, joints, and fittings also increase the frictional available-energy destruction. In this problem the majority of the friction cost ($\sim$ 90%) is a consequence of the valve, joint, and fitting friction

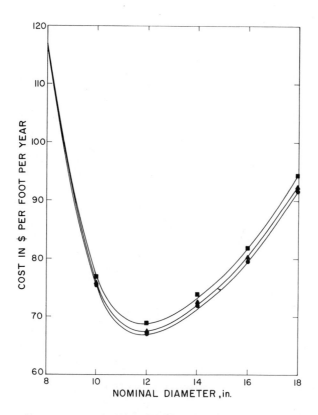

Figure 7.   Total cost curves as functions of nominal diameter for three insulation thicknesses: θ = 1.0 (■), 3.5 (●), and 7.0 (▲) in. (1 in. = 2.54 cm).

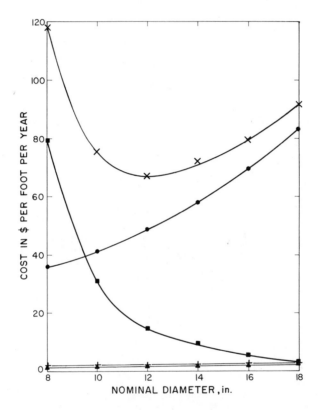

Figure 8.   Piping system costs as functions of nominal pipe diameter for θ = 3.5 in. (1 in. = 2.54 cm): (×), total cost; (●), pipe cost; (■), friction cost; (+), insulation cost; and (▲), heat loss cost.

rather than pipe friction. For piping systems with a smaller proportion of fittings, valves, etc.--compared to lengths of straight pipe--the heat loss and insulation costs will be more important.

The functional dependence of the total cost and the various operating and capital costs on the nominal pipe diameter are shown in Figures 7 and 8. Figure 8, while showing that heat-loss and insulation costs are insignificant (in this problem), also illustrates the fact that, as the investment in the pipe increases (larger D_n), the cost of the frictional available-energy destruction is reduced. For some pipe diameter, the total cost is a minimum--larger and smaller diameters yield higher costs. With the 18-inch piping system, the frictional available energy destruction is diminished, but the expensive capital investment is prohibitive. With the 8-inch system, the pipe cost is decreased, but the friction cost is unacceptably high. It is the 12-inch system that minimizes the sum of the pipe and friction costs with 3.5 inches of insulation.

The heat-loss and insulation costs are small compared to the friction and piping costs; only 5.6% ([$1.74/ft-yr + $2.14/ft-yr]/ $67.13/ft-yr) of the <u>total cost</u> for the optimal piping system. Furthermore, the effect of increased insulation thickness is not great. Figure 7, which shows total cost curves at selected insulation thicknesses (θ = 1.0, 3.5, and 7.0 in.), illustrates this effect. When the heat-loss and insulation costs are of secondary importance, the insulation thickness might well be dictated by a safe-surface temperature criterion.

It is interesting to note that the optimal insulation thickness happens to be the same for each pipe diameter analyzed in this problem. The insulation and heat-loss cost and, consequently, the optimal thickness are primarily functions of the steam temperature. Since, for each pipe diameter, the steam temperature was 1063 R, the optimal insulation thickness did not change. However, if the steam temperature is increased to 1163 R, the optimal insulation thickness becomes 4.0 inches. Although high-temperature applications (T_{st} > 1300 R) call for larger insulation thicknesses due to increased heat-loss costs, special piping materials are required, which makes the piping cost the single most significant factor, minimizing the significance of the increased heat-loss cost.

It is important not to make any general conclusions regarding the relative costs of piping, friction, insulation, and heat loss based solely on this example. Recall that the piping and friction costs for this problem are quite high and greatly overshadow the insulation and heat-loss costs. However, there are many situations in which the heat-loss and insulation costs are substantial fractions of the total cost. Consider a case where the piping is straight, with no fittings, expansion joints, etc., and such that its length equals the equivalent length of the piping and fittings in Figure 6, 49.7 ft. The straight pipe is carrying steam under

the same conditions as bleeder steam line, B5. The optimal nominal pipe diameter then turns out to be 14 inches, the optimal
insulation thickness is 3.5 inches, and the corresponding minimum
cost is $8.89/ft-yr. The breakdown of costs is as follows: fric-
tion cost, $1.33/ft-yr (15.0% of the total); pipe cost,
$.53/ft-yr (39.6%); heat-loss cost, $1.74/ft-yr (19.6%); and the
insulation cost, $2.30/ft-yr (25.8%).

Closure

The foregoing results illustrate three simple, practical applications of Second-Law costing methods: (i) cost accounting,
(ii) operation and maintenance decisions, and (iii) optimal design. The key is the assignment of costs to the true commodity of
value, available energy. This allows the appropriate costs to be
determined for the different commodities (such as steam, electric
charge, etc.) which carry available energy, at any juncture within
a complex system.
It is available energy that fuels the processes occurring in
any device. For any subsystem or component in a complex system,
knowledge of the cost of the available energy supplied to a device
allows an economic analysis of that device to be made, in isolation from other components of the system. Thus, for optimal design, maintenance, and operation purposes, decisions can be made
more simply, without contending with the whole system.   Nevertheless, it needs to be mentioned that the optimal design of a
system, by optimization of one unit at a time, is an iterative
process. Improved selection of one unit affects the unit costs
of available energy being supplied to other units and requires
re-optimization. Furthermore, when dealing with one unit at a
time, more insight into system operation is obtained; and, for
example, creativity is enhanced for improving the system via
process modification.
It is hoped that these illustrative examples will serve to
crystallize the principles surveyed by Reistad and Gaggioli (2),
and that together with their paper will motivate the readers to
apply available energy accounting--taking advantage of it for
their own practical purposes, while also advancing the state of
the art.

List of Symbols

$\dot{A}$ = available energy flow

$\dot{A}_{Bi}$ = available energy flow in the ith bleed line

$\dot{A}_{F,add} \equiv$ additional boiler fuel (available energy)

$\dot{A}_f$ = rate of available energy consumption due to friction

$\dot{A}_Q$ = rate of available energy loss by heat transfer

c	=	unit cost of available energy
c_{Bi}	=	unit cost of available energy in i^{th} bleed line
c_F	=	unit cost of fuel (available energy)
c_{st}	=	unit cost of steam available energy
D_n	=	nominal pipe diameter
D_P	=	pipe diameter
f	=	friction factor
g_c	=	gravitational constant
K_i	=	loss coefficient of ith joint or valve
L	=	length of piping system
$\dot{m}$	=	mass flow rate
$\dot{Q}$	=	rate of heat transfer
T_0	=	dead state (ambient) temperature
T_{st}	=	steam temperature
V	=	velocity
Z_I	=	capital cost of insulation
Z_P	=	capital cost of pipe
ϕ	=	thickness
ϕ_F	=	friction cost
ϕ_I	=	insulation cost
ϕ_P	=	pipe cost
ϕ_Q	=	heat loss cost
ϕ_T	=	total cost
ψ	=	amortization cost
$\dot{\$}$	=	money flow rate
Re	=	Reynolds number

Literature Cited

1. Gaggioli, R.; Wepfer, W. J., Paper No. 60A presented at the 85th Annual A.I.Ch.E. Meeting, June 7, 1978, Philadelphia, PA
2. Reistad, G.; Gaggioli, R.   "Available-Energy Costing," this volume.
3. Gaggioli, R.; Wepfer, W.; Chen. H.; Trans. A.S.M.E.--J. Eng. Power, 1978, 100, 511.

4. Wilson, W.; Kovacik, J.   Hydrocarbon Processing, December, 1976.
5. Gaggioli, R. Proc. Amer. Power Conf., 1975, 37, 656.
6. Fehring, T.; Gaggioli, R.   Trans. A.S.M.E.--J. Eng. Power, 1977, 99, 482.
7. Obert, E.; Gaggioli, R.   "Thermodynamics," 2nd ed., New York: McGraw-Hill, 1963.
8. Gyftopoulos, E.; Lazaridis, L.; Widmer, T.   "Potential Fuel Effectiveness in Industry," Cambridge: Balinger, 1974.
9. Potter, P., "Power Plant Theory and Design," 2nd ed., New York: Roland, 1959.
10. Skrotzki, B.; Vopat, W.   "Power Station Engineering and Economy," New York: McGraw-Hill, 1960.
11. "ASHRAE Handbook of Fundamentals," New York, 1972.
12. DeNevers, N. "Fluid Mechanics," Reading, Mass.: Addison Wesley, 1970.
13. Wepfer, W; Gaggioli, R.; Obert, E.   Trans. A.S.M.E. - J. Eng. Industry, 1979, 101 (in press).

RECEIVED November 1, 1979.

Economic Selection of a Venturi Scrubber

PETER J. PETIT

Coal Gasification Systems Operation, Allis–Chalmers Corporation,
Milwaukee, WI 53201

Selecting a vendor is a task which is frequently encountered
in the chemical process industries. Several quotations are sol-
icited from different vendors on the same system or piece of
equipment, with the intent of awarding the sale to the vendor
having the most attractive combination of performance and cost.
If performance does not vary significantly from quote to quote,
the task is reduced to determining which alternative represents
the lowest total (capital plus operating) cost.

There is usually little problem in obtaining capital costs
from the vendors, but operating costs are often more difficult to
establish. This is particularly true if the equipment utilizes
an unconventional form of "fuel" (such as steam, waste heat or a
pressure drop) instead of or in conjunction with the more common
fuels (such as coal, oil, or electricity). Not only can it be
more difficult to price accurately the unconventional fuels, but
also variations in the rate at which these fuels are used may have
significant impact on the flows and balances in the rest of the
plant, of which the equipment is to be a part.

The traditional means for assessing operating cost involves
the use of a perperty called energy, the obvious strategy being
to determine a "cost per unit energy" for each form of fuel, which
can then be applied to the flow rate of that fuel. Though this
strategy works well in the more straightforward cases, there is
one major flaw in its logic which can lead to errors in more complicated problems: Energy does not define a substance's value as
a fuel.

For example, a unit of electrical energy can potentially do
more "work" than a unit of low temperature steam energy, and a
unit of energy in air at ambient temperature, pressure and humidity
can do nothing. At the same price per unit energy, electricity is
the best bargain of the three.

What property, if not energy, imparts fuel value to a substance? The answer is available energy. Available energy is any
form of potential energy (thermal, chemical, electrical, etc.)
residing in a substance which is not in equilibrium with the

0–8412–0541–8/80/47–122–187$05.00/0
© 1980 American Chemical Society

environment, measuring the substance's capacity to drive pro-
cesses--to perform useful "work." ("Potential energy" is used
here in a broader sense than the traditional concept of energy
associated with a conservative force field.)

This paper addresses the problem of choosing between two vendors who have submitted quotations on a venturi scrubber and cyclone combination which will be used in a large plant to clean a
pressurized gas stream. The method of economic analysis employed
here is based on the concept of available energy, which is rooted
in the second law of thermodynamics, rather than on energy, which
is a first law concept.

When energy is used to assess operating costs in this type of
problem, it generally becomes necessary to carry out a detailed
energy balance to assess correctly the effect on plant operating
cost of switching from one vendor's equipment to another. In
large or complicated systems where this is impractical, corrections (efficiencies) are sometimes employed which put all energy
units on an "equivalent" basis. Unfortunately, this last method
is prone to error and misinterpretation.

In this paper, the property called available energy will be
used to assess operating costs. Because available energy does
define a substance's potential value as a fuel, no corrections are
needed when the consumption of one fuel is compared to that of another. For example, one unit of steam available energy serving
to "fuel" a turbine is fully equivalent to one unit of electrical
available energy delivered to an electric motor.

Extensive treatments on the meaning and methodology of available energy have been presented in several other papers (1-7).
The emphasis here will be placed on illustrating another example
of the application of available energy costing analysis to a
real-life problem.

Application of Available Energy Methodology to Vendor Selection Problem

The following is a case study in the application of second
law analysis to problems in which choices between alternative concepts must be made on the basis of overall economics.

Quotes from two vendors were obtained on a venturi scrubber
and cyclone combination, similar to that shown in Figure 1, to be
used for cleaning a pressurized gas stream. The information from
the quotes is contained in Table I. From the information presented it is not immediately clear which equipment would be the better choice. A superficial look at the table might lead to the
conclusion that Unit A is preferable: Its initial cost is less
and the operating costs associated with pump power will be less.
After further observation, however, it is realized that there is
an operating cost associated with the pressure drop of the gas
flowing through system A; "fuel" must be supplied somewhere in the
system in order to compensate for this. Should other matters be

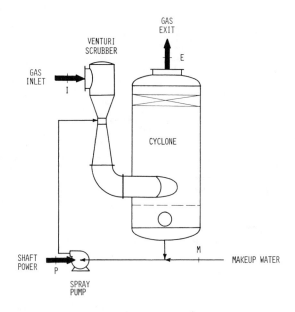

Figure 1.   Combination venturi scrubber and cyclone for gas cleaning

TABLE I.

INFORMATION CONTAINED IN VENDOR QUOTES ON
VENTURI SCRUBBER AND CYCLONE UNITS

	Vendor A		Vendor B	
System Cost	$8000		$9000	
Cleaning Efficiency	97.0%		97.0%	
First Law Efficiency	98.7%		98.9%	
Scrubber Pressure Drop	55.1 kPa	8 psi	0 kPa	0 psi
Spray Pump Power	0.15 kW	0.2 hp	3 kW	4 hp
Gas Flow Rate:	kg/hr	lb/hr	kg/hr	lb/hr
Inlet	24,755	54,576	24,755	54,576
Exit	24,765	54,598	24,753	54,572
Gas Temperature:	°C	°F	°C	°F
Inlet	37.8	100	37.8	100
Exit	36.1	97	37.8	100
Inlet Gas Pressure	474.3 kPa	68.8 psi	474.3 kPa	68.8 psi

pressing and the choices appear to be evenly matched, it might be hard to justify taking the time to work through the extent of analysis that a first law study (necessarily, of the whole system) would require; since the capital cost is substantially lower, system A would probably be chosen.

In contrast, it is hard to justify not using an available energy analysis in this case, since the logic and calculations, being straightforward, take so little time and effort.

An available energy analysis of this problem consists of first evaluating the rates of available energy destruction for the two systems. Using the notation of Figure 1, the same available energy balance can be written for both A and B:

$$\dot{A}_\delta = \text{Available Energy Destruction}$$
$$= (\dot{A}_I - \dot{A}_E) + \dot{A}_p + \dot{A}_M = \Delta\dot{A}_{gas} + \dot{A}_p + \dot{A}_M$$

Since the makeup water is essentially "free" from the environment it transports no available energy:

$$\dot{A}_M = 0 \text{ for both vendors A and B.}$$

The available energy input to the spray pump is merely the pump shaft power requirement:

$$\dot{A}_p = \dot{W}_{shaft} \text{ for both vendors A and B.}$$

Available energy transported with the gas will be evaluated as three independent contributions: thermal, pressure, and chemical. The change in total available energy content of the gas may be expressed

$$\Delta\dot{A}_{gas} = \dot{A}_I - \dot{A}_E = \dot{m}_I (a_{therm,I} + a_{press,I} + a_{chem,I})$$

$$+ \dot{m}_E (a_{therm,E} + a_{press,E} + a_{chem,E})$$

Note from the information in Table I that the change in mass (and presumably the change in composition) of the gas as it flows through either scrubber is very small. If we assume that in both cases $\dot{m}_E = \dot{m}_I$, then the above expression for $\Delta\dot{A}_{gas}$ reduces to:

$$\Delta\dot{A}_{gas} = \dot{m}_I (\Delta a_{therm} + \Delta a_{press} + \Delta a_{chem})$$

If the composition remains essentially unchanged throughout either scrubber, then

$$\Delta a_{chem} \doteq 0 \text{ for both vendors A and B.}$$

The change in thermal available energy per unit mass will be approximated using $c_p = 1 \text{ kJ/(kg K)} = .24 \text{ Btu/(lb }^\circ\text{F)}$, assumed constant for such a small change in temperature:

$$\Delta a_{therm} = \int_{T_I}^{T_E} c_p (1 - \frac{T_0}{T}) dt$$

$$\doteq c_p (T_E - T_I) - T_0 c_p \ln (T_E/T_I)$$

$$\Delta a_{therm} = 0 \text{ for vendor B, since } T_E = T_I.$$

$$\Delta a_{therm} = -.09 \text{ kJ/kg} = -.04 \text{ Btu/lb gas for vendor A.}$$

The drop in available energy content of the gas stream due to pressure reduction can be quickly evaluated:

$$\Delta a_{press} = RT_0 \ln (p_E/p_I)$$

For vendor B,

$$\Delta a_{press} = 0 \text{ since } P_E = P_I$$

But for vendor A,

$$\Delta a_{press} = 0.0083178 (\text{kJ/gmole }^\circ\text{K}) (23\text{g/gmole}) (298.15^\circ\text{K}) \times$$

$$\ln (419.2/474.3) = - 13.32 \frac{\text{kJ}}{\text{kg}} = -5.73 \text{ Btu/lb}$$

The change in total available energy content of the gas stream may now be evaluated by substituting the above values into the expression for $\Delta \dot{A}_{gas}$.

A summary of the destructions of available energy in the system is presented in Table II. An available energy consumption analysis lends itself to this method of presenting results since comparison of corresponding consumptions may be done at a glance.

TABLE II.

SUMMARY OF AVAILABLE ENERGY CONSUMPTIONS FOR
VENTURI SCRUBBER/CYCLONE COMPARISON

Location of Consumption of Available Energy	Vendor A		Vendor B	
	kw	hp	kw	hp
Pump Power, A_p	0.15	0.2	3	4
Makeup Water, A_M	0	0	0	0
Gas Losses, A_{gas}:				
thermal	0.7	0.9	0	0
chemical	0	0	0	0
pressure	91.6	122.9	0	0
Total Consumption	92.45	124.0	3	4

Conclusions

Presuming that the information quoted by the vendors is accurate, then scrubber system A is effectively consuming over 30 times the "power" that system B consumes to do the same job! The results, presented in Table II, show clearly that the 55.1 kPa (8 psi) pressure drop reduces all other consumptions of available energy to insignificance. If additional compressor capacity must be purchased to cover this power drain, the added capital cost would be roughly

$$\$ = \frac{123}{\eta_{II}} (\$50/hp) = \$9276$$

where $\eta_{II} = 0.68$ represents a typical second law efficiency for a compressor. (Compressor costs are usually scaled on input power, which may be estimated here using as assumed value of η_{II}. The value of η_{II} used here neglects any thermal available energy which the compressor might impart to the gas. This was done so that an accurate estimate of additional compressor input power might be obtained from the pressure available energy rise alone.) This raises the effective capital cost of choosing vendor A to \$17,276.

Furthermore, assuming 300 hrs/yr operation and a compressor with η_{II} = 0.68, the cost of power to operate system A would be

$/yr = (.15 + 91.6/0.68) kW ($0.025/kW-hr)(3000 hrs/yr)

   = $10,114/yr

compared to system B operating costs of

$/yr = 3 kW ($0.025/kW-hr)(3000 hrs/yr) = $225/yr

Under such circumstances, the scrubber/cyclone of vendor B is clearly the better choice. Since no additional compressor capacity would be required, the actual capital costs associated with vendor B are nearly 50% lower than for vendor A. Furthermore, vendor B's operating costs due to power consumption are almost negligible compared with vendor A's--lower by a factor of 45, when a typical compressor efficiency (η_{II}) is taken into account. (With such evidence in hand, one might want to give vendor A a change to modify his operating specifications--or his price!)
   (In this example, a subtle advantage of second law analysis has come to light. Any available energy flow or consumption can properly be expressed directly in kW (hp), which often tends to improve one's "feel" for what is actually happening. In contrast, it is usually inappropriate to express certain kinds of energy flows or losses in dimensional units normally reserved for work or electricity.)

Closure

For the venturi scrubber/cyclone comparison, evaluation of available energy consumptions revealed that the 55.1 kPa (8 psi) pressure drop of system A implied additional operating and capital costs which by far exceed its apparent cost savings over system B. This dramatic revelation proceeded from only a very few straightforward calculations.
   The available energy consumption represented by the pressure drop was 92 kW (123 hp). These results not only illuminated the impact of the pressure drop on power consumption but also enabled the associated capital and operating cost implications to be quickly and straightforwardly evaluated.

List of Symbols

$\dot{A}$ = available energy per unit time
a   = specific available energy
c_p = heat capacity at constant pressure
h   = specific enthalpy
$\dot{m}$ = mass flow rate
p   = pressure

R_0 = universal gas constant
s   = specific entropy
T   = temperature
$\dot{W}$ = power; work rate
x   = mole fraction
η_I = first law efficiency
η_{II} = second law efficiency; true efficiency

Subscripts and Superscripts

δ = destruction
E   = exit
I   = inlet
M   = makeup water
P   = shaft power
0   = reference ("dead") state

Literature Cited

1. Gaggioli, R.; Petit, P. Chemtech, 1977, 7, 496.
2. Rodriguez, L.; Petit, P. "Calculation of Available Energy Quantities," this volume.
3. Wepfer, W.; Gaggioli, R.; Obert, E. "Proper Evaluation of Available Energy for HVAC," ASHRAE Paper No. 2524 (to appear in ASHRAE Transactions, 1979).
4. Tribus, M.; Evans, R. "Thermoeconomics," UCLA Report No. 52-63, 1962.
5. Obert, E.; Gaggioli, R. "Thermodynamics," 2nd ed., New York: McGraw-Hill Book Co., 1963.
6. Reistad, G.; Gaggioli, R. "Available Energy Costing," this volume.
7. Wepfer, W. "Applications of Availability Accounting," this volume.

RECEIVED October 26, 1979.

12

Economic Selection of the Components of an Air Separation Process

MANSON BENEDICT and ELIAS P. GYFTOPOULOS

Massachusetts Institute of Technology, 77 Massachusetts Ave., Cambridge, MA 02139

This paper presents an availability analysis of one type of
oxygen production cycle centering around separation of oxygen and
nitrogen in a fractionating tower. The plant is driven by work
inputs to compressors and blowers. The analysis shows the irreversible entropy production in the various units and, in turn, the
added work inputs required as a consequence thereof. Furthermore,
a comparison is made with an ideal process of the same type,
wherein all irreversibilities are reduced to the minimum possible,
subject to the constraints imposed by (a) the use of a tower, and
(b) the properties of the flowing streams.

In addition, it is shown how the total cost of oxygen production varies with the irreversibility in the main heat exchanger.
By relating this cost to the design parameters of the heat exchanger (pressure drop and warm-end temperature difference), the
parameter values that minimize the total cost are determined.
Thus, with the main heat exchanger as an example, it is shown how
availability methods can be employed in design for "optimal"
selection of components.

This work was first presented in an unpublished lecture given
at MIT in 1949, along with similar analyses of several other chemical processes.

Oxygen Separation Process

This paper is concerned with the amount of work needed to
separate air into an oxygen-rich fraction containing 90 mole percent oxygen and a nitrogen-rich fraction containing 99 mole percent nitrogen. The minimum work required in a thermodynamically
reversible process conducted in an environment at one atmosphere
and 77°F with feed and product gases at these conditions is 421
Btu per pound mole of air fed.

The amount of work used in a practical oxygen separation
process is much greater than this minimum because of irreversibility. Figure 1 illustrates the process flow for one type of oxygen
production cycle. The process quantities refer to separation of

0–8412–0541–8/80/47–122–195$05.00/0
© 1980 American Chemical Society

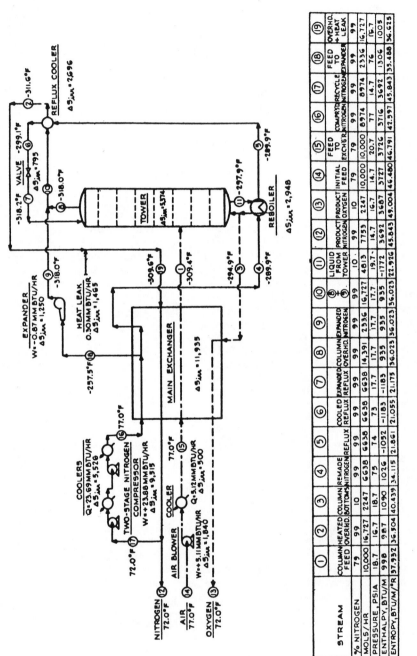

Figure 1. Practical oxygen process

10,000 pound moles of air per hour into a waste fraction containing 99 percent nitrogen and a product fraction containing 90 percent oxygen. The plant produces 380 tons of oxygen per day with
a recovery of 96 percent.

Air at 77°F is compressed by a blower to a pressure of 20.7
psi, which is sufficient to force the air through the exchangers
and towers of the separation plant. In the main exchanger, the
air is cooled by means of the outgoing product oxygen and nitrogen,
and emerges at its dewpoint at -309.4°F. In the tower, the air is
fractionated into the nitrogen overhead and oxygen bottom streams.
These streams are returned through the main exchanger and discharged from the plant at 72°F. The nitrogen is also used to sub-
cool tower reflux and is finally discharged at atmospheric pressure. The oxygen, which emerges from the tower at a somewhat
higher pressure than the nitrogen, is discharged from the plant at
16.7 psi.

To reboil the tower, to provide reflux, and to satisfy the
refrigeration requirements of the plant, an auxiliary nitrogen
stream amounting to 90 percent of the air fed to the plant is compressed in a two-stage compressor to 77 psi and is also cooled to
its dewpoint of -289.9°F in the main exchanger. This nitrogen is
liquefied in the reboiler, where its latent heat reboils the tower,
is subcooled against outgoing product nitrogen in the reflux
cooler, and finally is flashed through a valve into the top of the
tower where it provides reflux at -318.2°F. In the tower, the
nitrogen reflux stream is vaporized into the product nitrogen and
flows with it through the nitrogen pass of the main exchanger
where it gives up its heat to the incoming compressed nitrogen.
To compensate for the heat leak to the plant and the enthalpy difference between the outgoing nitrogen and oxygen streams and incoming air, a portion of the compressed nitrogen is made to do
work in the expander in order to lower its temperature from
-275.5°F to -318.1°F.

In setting up the conditions for this process, the following
assumptions have been made: (a) pressure drops of 2 psi through
the main exchanger and 1 psi through the other exchangers and from
top to bottom of the tower; (b) a temperature difference of 5°F
for each of the gas-to-gas exchangers; (c) a net heat upflow in
the tower equal to 3.5 percent greater than the minimum flow
needed to carry out the separation with an infinite number of
plates; (d) a heat leak of 30 Btu/mole of incoming air or 300,000
Btu/hr; and, (e) efficiencies for the compressors, the blower, and
the expander equal to 75 percent. Thus, a total of 29 x 10^6 Btu/
hr of work is expended in the nitrogen compressors and the blower,
and 0.87 x 10^6 Btu/hr of work is recovered in the expander. The
net work input is 28 x 10^6 Btu/hr or 10.8 kWh per 1000 ft^3 of oxygen. The theoretical work of carrying out this separation, evaluated from the change in enthalpy and entropy of the feed and products, is only 4.6 million Btu/hr or 1.7 kWh per 1000 ft^3 of oxygen.

The process inefficiencies that are responsible for the increased work expenditure can be identified by evaluating the entropy production of each element of the process, namely, by evaluating the irreversibilities. The largest portions of this entropy production, shown in Figure 1, are the main exchanger, the nitrogen compressors, the nitrogen coolers, and the tower.

To improve the effectiveness of the process, it is necessary to determine the extent to which the entropy production is characteristic of the process and the extent to which it is a consequence of equipment inefficiencies that could be reduced by increased capital expenditure for either larger or more effective equipment. To do this, we consider an idealized process design for a plant in which the exchangers are so large that the pressure drop is zero and the temperature difference is the minimum consistent with process heat balance. In addition, we assume that the heat leak may be made negligible by elaborate insulation, that only the minimum heat needed to reboil the tower will be used, and that the nitrogen compressor may be made 100 percent efficient and isothermal.

Figure 2 shows the flow sheet for such an idealized process. The work input is reduced from 28×10^6 Btu/hr to 7.7×10^6 Btu/hr, and the irreversible entropy production has been made practically negligible in all pieces of equipment except the tower. The entropy production here is still high because of the fact that fractionation occurs away from equilibrium conditions at all points of the tower except at the feed point.

Table I compares the work input and rate of entropy production in the practical process with corresponding quantities in the ideal process. This brings out the fact that the single place where the greatest reduction in power can be effected is in the main exchanger, and also shows the relative importance of the individual pieces of equipment as contributors of the total work input. It also shows that the total work input is the sum of (a) the theoretical work, and (b) the work equivalent of irreversible entropy production, given by the product of ΔS_{irr} and ambient temperature T_0.

To illustrate how the entropy production may be used to select approximate optimum design conditions for individual units of process equipment, we calculate the effect of varying the parameters of the main exchanger upon the total cost of the exchanger, the compressors, and the power required to run the plant for 5 years. The results are given in Figure 3. The figure illustrates the cost balance on this exchanger. The pressure drop and the warm-end temperature difference for which the exchanger is designed contribute to the cost of the plant in two principal ways. First, the lower the pressure drop and temperature difference, the larger the exchanger and the greater its cost. Second, the lower the pressure drop and warm-end temperature difference, the lower the irreversible rate of entropy production in the main exchanger, the smaller the compressors needed for the plant and the lower the

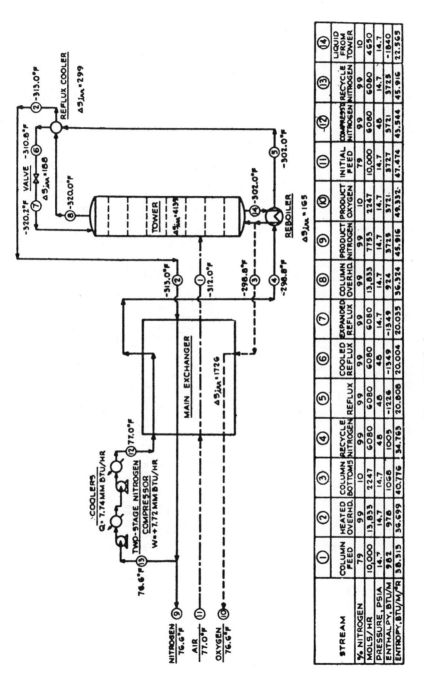

Figure 2.   Ideal oxygen process

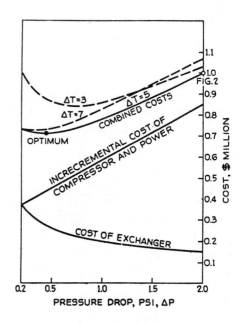

EXCHANGER SURFACE:

$A = 1.886 \times 10^6 \, \Delta P^{-0.4} \, \Delta T^{-1.4}$

POWER:

$\Delta S_{irr} = 3165 \Delta P + 776 \Delta T$

$KW = 0.27 \Delta S_{irr}$

UNIT COSTS:

SURFACE, $1/SQ.FT..

COMPRESSOR + 5 YEAR

POWER, $252/KW

Figure 3. Cost balance on main ex-changer

Table I.
Entropy Production Breakdown in Oxygen Processes
(Feed Rate: 10,000 Moles Air/hr)

PROCESS	IDEAL		PRACTICAL	
	Work Input, 10^6 Btu/hr	$T_0\Delta S_{irr}$, 10^6 Btu/hr	Work Input, 10^6 Btu/hr	$T_0\Delta S_{irr}$, 10^6 Btu/hr
Air Blower		–	5.11	0.988
Air Cooler		–		0.268
N$_2$ Compressor	7.72	–	23.88	5.109
N$_2$ Coolers		–		2.967
Main Exchanger		0.927		6.409
Expander		–	-0.87	0.671
Tower		2.222		2.886
Reboiler		0.089		1.583
Reflux Cooler		0.161		1.448
Valve		0.101		0.427
Heat Leak		–		0.787
TOTAL	7.72	3.50	28.12	23.54
$T_0\Delta S_{irr}$, 10^6 Btu/hr		3.50		23.54
Theoretical Work=$\Delta H - T_0\Delta S$		4.21		4.61*
Predicted Work, 10^6 Btu/hr		7.71		28.15

*The 4.61 consists of the 4.21 units of ideal theoretical work plus 0.40
unit needed because the N$_2$ and O$_2$ leave at T = 72°F < 76.6°F; i.e., the
0.40 unit is the theoretical work done by cooling from 76.6°F to 72°F.

cost of power supplied to the plant. The equation given in the
figure indicates how the exchanger surface (A) will vary with
pressure drop and temperature difference. The assumed cost for
exchanger surface is $1/ft^2.

The effect of pressure drop and temperature difference in increasing the rate of entropy production from the ideal case to the
practical case is represented by the equation given in Figure 3.
Were the availability efficiency 100 percent for the compressors,
their coolers, and motors, then the additional power required because of irreversibility in the heat exchanger would simply be
kW = $T_0\Delta S_{irr}$ = 537°R $[\Delta S_{irr}(Btu/°R\text{-}hr)]$ $[Btu/3412 \text{ kWh}]$ = 0.157
$[\Delta S_{irr}(Btu/°R\text{-}hr)]$. Assuming a composite availability efficiency
of 58 percent for compressor and auxiliaries, the actual power requirement would be kW = 0.27 ΔS_{irr}. It is assumed that the initial
cost of the compressors is increased by $50 for every kilowatt in-
crease in power input and that power is to be charged at the rate
of 0.6¢/kWh for a period of 5 years. Assuming an interest rate of
5.5 percent and a 90-percent plant capacity factor, the present
worth of the total cost of power is $252 per kilowatt or $68 per
unit rate of entropy production. (As this economic analysis was
made in 1949, costs and interest rate are no longer representative.

Electricity has increased by a factor of about 4, and capital equipment and interest rate by a factor of about 2.)

The solid curves in Figure 3 show the effect of pressure drop upon cost for a warm-end temperature difference of 5°F. The opposing tendencies of entropy production and exchange surface are indicated. Total cost is minimized at a pressure drop of about 0.5 psi rather than the 2 psi used in the practical process. Similar curves for the total cost at temperature difference 3°F and 7°F indicate that a warm-end temperature of 5°F is close to the optimum.

Thus, by consideration of the rate of entropy production in the main exchanger, the optimum pressure drop and temperature difference across this exchanger may be estimated without the necessity for a complete system design at each combination of conditions.

Closure

In a real application, dating back to 1949, this paper concisely illustrates the theme of the symposium: the use of availability (a) to calculate "work penalties" for the various irreversibilities in a process, and (b) to determine the total costs attributable to the irreversibilities and the use of these costs for "optimal" design. Another important result is the determination of the minimum work for a process with unavoidable irreversibilities, such as those associated with the fractionating tower here.

Editor's Note:

This paper illustrates the use of entropy balances in Second-Law analysis. See the editor's note at the end of the article by Cremer, in this volume.

RECEIVED October 29, 1979.

MODERN THERMODYNAMICS

A Thermodynamic Theory for Nonequilibrium Processes

RICHARD A. GAGGIOLI

Department of Mechanical Engineering, Marquette University, 1515 W. Wisconsin Ave., Milwaukee, WI 53233

WILLIAM B. SCHOLTEN

Intertechnology, Inc., Warrenton, VA 22186

Traditional developments of thermodynamics defined entropy – and hence thermostatic property relations – only for equilibrium states. This is because entropy has been defined only for states that can be connected reversibly to some common reference state. In the present approach, this restriction is overcome by eliminating reversibility in favor of a more general concept – called ideality - which includes reversibility as a special case.

The key to the generalization is to base the Second Law on the property availability (1,2,3); the availability at any state of a system reflects the extent to which the system can affect any other because the state is not in equilibrium with some reference datum.

A process is said to be ideal if there is no destruction of this capacity to affect other systems; that is, if there is no destruction of availability. Not every ideal process is reversible; for example, if a given state is not an equilibrium state (internally) it cannot be a state of a reversible process. However, it can be a state of an ideal process. In particular, it is the initial state of any process which would furnish its availability; (i.e., any process which goes from the given state to one in equilibrium with the reference, and does indeed have the maximum effect on another system). Thus, the concept of ideality can be used to replace the concept of reversibility as a more general criterion for the perfection of processes.

Another feature of the present theory is that it provides a formalism for deducing a complete mathematical representation of a phenomenon. Such a representation consists, typically, of (1) Balance equations for extensive properties (such as the "equations of change" for mass, energy and entropy); (2) Thermokinematic functions of state (such as pv = RT, for simple perfect gases); (3) Thermokinetic functions of state (such as the Fourier heat conduction equation $\underline{q} = -k(T,p)\nabla T$); and (4) The auxiliary conditions (i.e., boundary and/or initial conditions). The balances are pertinent to all problems covered by the theory, although their formulation may differ from one problem to another. Any set of

0–8412–0541–8/80/47–122–205$05.50/0
© 1980 American Chemical Society

functions of state (constitutive relations) pertains to a class of problems; the classes are distinguished from one another by constitution and by the realm of states encompassed. The balance equations and functions of state are called the governing equations of the class of problems. In any class of problems defined by a set of functions of state, the problems are distinguished from one another by the auxiliary conditions; these reflect the conditions of the entity at a given time, and the influences imposed upon it from outside. Then, when complemented by appropriate auxiliary conditions, a set of governing equations defines completely the reaction of the entity, in the given configuration, when subjected to those outside influences, and hence the solution of the system portrays the resultant behavior of the entity.

The key to implementing this generalization from reversibility to ideality is the introduction of other integrating factors in addition to temperature. Traditionally, the Gibbs equation (e. g., du = Tds - pdv for simple states) is derived, via reversible processes, by invoking the integrating factor, temperature. Thus, by introducing this integrating factor the divergence of the heat flux vector can be replaced by the "time" derivative of a property. Thereby, the goal of deriving an equation where the only derivatives present are "time" derivatives of properties is reached. However, to go beyond the traditional and consider non-equilibrium states, it is necessary to eliminate not only the heat flux, but also certain local production rate functions. For example, to derive the Gibbs equation for chemical nonequilibrium states, via ideal processes, integrating factors are introduced not only to eliminate the heat flux, but also to eliminate the production rate of each of the reaction products.

Foundations

(Here only the essential concepts are highlighted. For a more complete presentation, including "physical" motivations and definitions of terms, see (2)).

Suppose that it is desired to portray quantitatively some real phenomenon; that is, it is desired to portray the phenomenon with a mathematical model. Variables are to be employed, then, to represent characteristics displayed by the actual objects involved in the phenomenon, to represent spontaneous changes which occur within the objects, and to represent interactions they have with one another. The variables to be called properties can be viewed as those employed to represent the various configurations of an object. Each property to be called extensive can be viewed as representing a content of an object; that is, the value of an extensive property at an instant represents how much of it is contained in an object. Variables will be defined to represent changes in extensive properties from two causes; transports and productions (including destructions as negative productions). The transport functions, then, may be viewed as portraying exchanges between objects;

hence, the transports represent the interactions between objects. Other changes in extensive properties, attributed to productions, then represent the spontaneous changes within an object. An extensive property of which there can be no production or destruction is said to be conserved.

Since it would often be desirable to measure quantitatively the mutual influence that objects have had, or will have had, upon one another, the invention of a means for measuring the extent to which objects interact would be desirable. This desirability has led to the concept of energy. The extent of the mutual influence which two objects have on one another is quantitatively measured by an amount of energy transferred between them during the interaction. Necessarily, energy must be a conserved content; i.e., it must be possible to change the energy content of an object only by transporting it in or out, and never by producing or destroying it within. For, if it could be produced or destroyed, the energy of an object could change without interacting with other objects.

These "physical" criteria are formalized in the following postulate, which is in essence a statement of the "First Law":

   Postulate I (the First Law). There exists an extensive property, to be called energy, which
         (a) is transported with every extensive property of an independent set.
         (b) is conserved.

In addition to having a means for quantitatively describing the extent of an interaction, it is also desirable to have a measure for describing the capacity of an object for influencing any other object. This leads to the concept of availability, which is the maximum amount of energy that could be transmitted to any other system; this capacity of a given system to affect other systems exists whenever the given system is not at stable equilibrium. (The reference datum is not simply a kinematic reference frame; the datum must provide a reference for the non-mechanical properties as well, and hence might well be called a reference atmosphere. When applying the theory, the reference datum selected must be an object which is not influenced by the phenomenon being analyzed (just as in Mechanics, where an inertial reference frame must be selected. See (7) for further elaboration on availability and reference datums.)

From this it is seen that the availability of an object can never be negative because an object could always transmit zero energy by not interacting at all. Furthermore, the capacity to influence other systems can be lost -- destroyed -- sometimes completely, sometimes partially. The various "physical" aspects of availability are formalized in another postulate, which is in essence a statement of the "Second Law."

   Postulate II (the Second Law). There exists a non-negative extensive property, to be called availability, which

(a) may be transported with every extensive property of
an independent set.

(b) cannot be created.

A real process will occur whenever it would continuously destroy
availability; it will extremize the availability destruction.   In
a hypothetical ideal process availability is conserved.

   In addition to energy and availability, constraints are included to reflect the possible degree of control from outside the
system. A constraint, then, is any extensive property besides energy or availability (or one defined as a function thereof). If a
constraint is conserved, it is called an external constraint (because its value for the system can be completely fixed from without); if it is not conserved it is called an internal constraint
(because its value can be somewhat affected but not completely
fixed from without). (It is the internal constraints that call for
the extra integrating factors discussed previously; one integrating
factor will be introduced for each independent internal constraint)
For example, in a given model the number of oxygen atoms may be an
external constraint, while the number of oxygen molecules would be
internal, provided the model incorporates production and destruc-
tion thereof via reactions such as $2H_2 + O_2 \leftrightarrow 2H_2O$, which would not
change the number of oxygen atoms.

Governing Equations for a Simple Compressible Flow

   A continuum flow (or deformation) is said, here, to be simple
if there is only one scalar constraint and it is conserved.

   To facilitate the discussion of balance equations, each constraint will be modeled by a three dimensional Euclidian manifold.
(Here "manifold" can be taken as a synonym for "space" as in (4,
pg 9).) Using this scheme, processes are modeled by employing a
parametric transformation between each manifold and the "manifold
of observation," or reference manifold. The parameter, to be denoted by σ , is called the process parameter; since a process occurs as σ varies continuously from one value to another. (For a
model of a "real" process, σ would normally be taken as time, t.)

   Here, the reference manifold will be called the volume manifold, and the manifold for the single scalar constraint will be
called the mass manifold. The word place will be used to mean
point in the volume manifold. Then the balance equation (or "local equation of change") for an extensive property of a continuum
is an expression for the rate of change of the amount of that property per unit of volume at a particular place. (In these continuum models, balance equations can be integrated over a region R in
the volume manifold to yield a balance for the region, which equates
the rate of change of the amount of the extensive property in R to
the sum of a net rate of transport function and a net rate of production function. In applying the theory to Lumped Parameter models, these region balances are written directly without recourse to

the local equations of change. See (2) for further elaboration.)
Thus, the equation of change for mass is

$$\frac{\partial \rho}{\partial \sigma} = - \nabla \cdot [\rho \underline{V}] \tag{1}$$

where ρ is the (mass) density (i.e., the amount of the mass manifold "constraint" per unit amount of the volume manifold
and $\rho \underline{V}$ is the mass flux. $\underline{V}$ is called the (mass) velocity and reflects the transport of the mass manifold through the reference
manifold.

Similarly, the local equation of change for energy is written
as

$$\frac{\partial (\rho e)}{\partial \sigma} = - \nabla \cdot [\rho e \underline{W}] \tag{2}$$

where e is the energy per unit mass, and $\underline{W}$ is the <u>energy velocity</u>.
    The local equation of change for availability follows the
same pattern as those for energy and mass, with the addition of a
production rate per unit volume term a_p. Thus

$$\frac{\partial (\rho a)}{\partial \sigma} = - \nabla \cdot [\rho a \underline{Z}] + a_p \tag{3}$$

where a is the availability per unit mass, and $\underline{Z}$ is the <u>availabil-
ity velocity</u>.
    Because the equations of change will be complemented by con-
stitutive relations, it is advantageous to re-write the balance
equations to reflect changes at a "particle of matter," rather than
at a "place in space." This is accomplished by using the "hydro-
dynamic" or "substantial" derivative. (Nevertheless, the tensor
nature of the quantities in these equations is such that they are
"volume tensors" and not "mass tensors." See (5) for elaboration.)
Thus,

$$\rho \frac{Dv}{D\sigma} = \nabla \cdot \underline{V} \tag{4}$$

where

$$v \equiv 1/\rho \tag{5}$$

$$\rho \frac{De}{D\sigma} = - \nabla \cdot \{\rho e [\underline{W} - \underline{V}]\} \tag{6}$$

$$\rho \frac{Da}{D\sigma} = - \nabla \cdot \{\rho a [\underline{Z} - \underline{V}]\} + a_p \tag{7}$$

Note that $[\underline{W}-\underline{V}]$ and $[\underline{Z}-\underline{V}]$ reflect the rates at which energy and availability respectively are moving relative to the mass manifold.

By Postulate I there is an energy transport associated with mass transport. By definition, $\underline{V}$ reflects the transport of the mass manifold through the volume manifold. Hence, the rate of energy transport with mass transport through the volume manifold is given by $\underline{\sigma}\cdot\underline{V}$, where $\underline{\sigma}$ is a second order tensor. Then

$$\rho e\underline{W} = \underline{\sigma}\cdot\underline{V} + \underline{q} \tag{8}$$

defines $\underline{q}$ -- which can be called the "heat flux" -- as any energy flux over and above that associated with motion of the mass manifold. Furthermore, it is customary to arbitrarily break down the portion $\underline{\sigma}\cdot\underline{V}$ into two parts: (1) a part representing energy carried in the mass -- i.e., the energy per unit mass e times the mass flux $\rho\underline{V}$ -- plus (2) a part for the energy flux (associated with the flow) relative to the mass flux, denoted as $\underline{\pi}\cdot\underline{V}$ -- which can be called the "work flux." This is accomplished without introducing additional variables (i.e., the variable $\underline{\sigma}$ is replaced by $\underline{\pi}$), by defining $\underline{\pi}$ as

$$\underline{\pi} \equiv \underline{\sigma} - \rho e\underline{g} \tag{9}$$

where $\underline{g}$ is the fundamental tensor; $\underline{\pi}$ is called the stress tensor. Hence

$$\underline{\sigma}\cdot\underline{V} = \underline{\pi}\cdot\underline{V} + \rho e\underline{V} \tag{10}$$

The expression for the energy flux for all simple flows is then

$$\rho e\underline{W} = \rho e\underline{V} + \underline{\pi}\cdot\underline{V} + \underline{q} \tag{11}$$

and the equation of change for energy for all simple flows is

$$\rho\,\frac{De}{D\sigma} = -\,\nabla\cdot[\underline{\pi}\cdot\underline{V} + \underline{q}] \tag{12}$$

Similar to Eq. 8, the availability flux can be expressed as a portion associated with the mass transport, $\underline{\omega}\cdot\underline{V}$, plus an "excess" $\underline{r}$ i.e.,

$$\rho a\underline{Z} = \underline{\omega}\cdot\underline{V} + \underline{r} \tag{13}$$

As before, $\underline{\omega}$ can be broken up into two parts, one of which represents the availability carried in the mass -- $\rho a\underline{V}$

$$\underline{\Xi} \equiv \underline{\omega} - \rho a\underline{g} \tag{14}$$

Furthermore, inasmuch as the availability contained by a volume manifold is the portion of the energy content of the manifold which could be transmitted out of the manifold were the "contained" manifolds brought to complete stable equilibrium, then the work flux of availability (through a particle) will be related to the work flux of energy by

$$\underline{\underline{\Xi}} = \underline{\underline{\pi}} - \underline{\underline{\pi}}_0 \tag{15}$$

where $\underline{\underline{\pi}}_0$ is the stress tensor (at the particle) in the hypothetical complete equilibrium state. (Equation 15 may be explained (1,2,3) with the physical argument that $\underline{\underline{\pi}}_0 = p_0\underline{\underline{g}}$ where p_0 is the normal compressive stress in the reference atmosphere -- atmospheric pressure. Then $p_0\underline{V}$ is the portion, of the "work flux" of energy, which would be delivered to the atmosphere; $[\underline{\underline{\pi}} - p_0\underline{\underline{g}}]\cdot\underline{V}$ -- the energy flux over and above that delivered to the atmosphere- - can be supplied to any other system (i.e., it is totally available).)

Thus the equation of change of availability for simple flows is written as

$$\rho\,\frac{Da}{D\sigma} = - \nabla\cdot[\underline{\underline{\pi}}\cdot\underline{V} - \underline{\underline{\pi}}_0\cdot\underline{V} + \underline{r}] + a_p \tag{16}$$

Next, the thermokinematic equations of change will be obtained. These are found by considering ideal relaxation processes, i.e., hypothetical *ideal* processes from any and every state visited during the real process to an arbitrary dead reference state. The foregoing equations of change are combined, for such a process, to eliminate all terms without parameter derivatives of properties; thence, the existence of functions relating the properties (i.e., the "state principle") can be deduced.

In physical terms, a simple compressible fluid is one which cannot exert shear stresses without viscous dissipation. In theoretical terms, it is defined as a fluid wherein, during *ideal* processes, all energy and availability transports with mass transports are collinear with $\underline{V}$; i.e. $\underline{\underline{\sigma}} = \sigma\underline{\underline{g}}$ and $\underline{\underline{\omega}} = \omega\underline{\underline{g}}$ so that $\underline{\underline{\sigma}}\cdot\underline{V} = \sigma\underline{V}$ and $\underline{\underline{\omega}}\cdot\underline{V} = \omega\underline{V}$ (and $\underline{\underline{\pi}}\cdot\underline{V} = p\underline{V}$ and $\underline{\underline{\Xi}}\cdot\underline{V} = [p - p_0]\underline{V}$). Then, for ideal relaxation processes Eqs. 12 and 16 become, respectively

$$\rho\,\frac{De}{D\sigma} = - \nabla\cdot[p\underline{V} + \underline{\tilde{q}}] \tag{17}$$

and

$$\rho\,\frac{Da}{D\sigma} = - \nabla\cdot[\{p - p_0\}\underline{V} + \underline{\tilde{r}}] \tag{18}$$

where the symbols p, $\underline{\tilde{q}}$ and $\underline{\tilde{r}}$ denote ideal process variables. Using Eq. 4, Eqs. 17 and 18 can be rewritten as

$$\rho \frac{De}{D\sigma} = - p\rho \frac{Dv}{D\sigma} - \underline{V} \cdot \nabla p - \nabla \cdot \tilde{\underline{q}} \tag{19}$$

and

$$\rho \frac{Da}{D\sigma} = - [p - p_0] \frac{Dv}{D\sigma} - \underline{V} \cdot \nabla [p - p_0] - \nabla \cdot \tilde{\underline{r}} \tag{20}$$

respectively.

Thus, two independent equations remain, but with three terms which have no parameter derivatives of properties. Note that the terms $\underline{V} \cdot \nabla p$ can be eliminated by restricting consideration to processes with $\underline{V} \equiv 0$, and that the terms $\underline{V} \cdot \tilde{\underline{q}}$ and $\underline{V} \cdot \tilde{\underline{r}}$ can be eliminated by consideration of processes with $\tilde{\underline{q}} \equiv 0$. During any ideal process with $\tilde{q} \equiv 0$, the quantity

$$s \equiv e + p_0 v - a \tag{21}$$

is constant. For, combining Eqs. 4, 12, and 16 yields

$$\rho \frac{Ds}{D\sigma} = - \nabla \cdot [\underline{q} - \underline{r}] - a_p \tag{22}$$

and the right hand side is zero for ideal processes that are adiabatic -- i.e., which have $\tilde{\underline{q}} \equiv 0$. (If $\underline{r} \neq 0$, then $\underline{q} \neq 0$, by Postulate I(a).)

It remains, now, to combine Eqs. 19, 20 and 22 for ideal processes in such a way to eliminate $\tilde{\underline{q}}$ and $\tilde{\underline{r}}$. For ideal processes, define α by

$$\alpha \equiv \frac{\nabla \cdot \tilde{\underline{r}}}{\nabla \cdot \tilde{\underline{q}}} \tag{23}$$

Then from Eq. 22, for ideal processes,

$$\rho \frac{Ds}{D\sigma} = - [1 - \alpha] \nabla \cdot \tilde{\underline{q}} \tag{24}$$

Thus, Eqs. 19 and 20 can be rewritten as

$$\rho \frac{De}{D\sigma} = - p\rho \frac{Dv}{D\sigma} - \underline{V} \cdot \nabla p + \left. \frac{\rho}{1-\alpha} \frac{Ds}{D\sigma} \right]_{ideal} \tag{25}$$

$$\rho \frac{D(a-p_0 v)}{D\sigma} = - p\rho \frac{Dv}{D\sigma} - \underline{V} \cdot \nabla [p - p_0] + \left. \frac{\rho\alpha}{1-\alpha} \frac{Ds}{D\sigma} \right]_{ideal} \tag{26}$$

Consider first all those particle states that can be connected to the dead reference state while $\underline{V} \equiv 0$. Then, for such ideal processes, the terms $\underline{V} \cdot \nabla p$ and $V \cdot \nabla p_0$ are zero in Eqs. 25 and 26. The derivatives in these equations are derivatives at a fixed particle. Inasmuch as e,v, and s are properties, the change in value of each, from any one of the states at hand to the dead reference state is independent of the intermediate states visited. Hence it follows from Eqs. 25 and 26, with $\underline{V} \equiv 0$, that there exist functions

$$e = u(s,v) \Big]_{\underline{V} \equiv 0} \tag{27}$$

$$a - p_0 v = b(s,v) \Big]_{\underline{V} \equiv 0} \tag{28}$$

It also follows that, for these states,

$$p = -\frac{\partial u}{\partial v}(s,v) \qquad\qquad \frac{1}{1-\alpha} = \frac{\partial u}{\partial s}(s,v) \tag{29}$$

$$p = -\frac{\partial b}{\partial v}(s,v) \qquad\qquad \frac{\alpha}{1-\alpha} = \frac{\partial b}{\partial s}(s,v) \tag{30}$$

Note that p and α are thus functions of properties, so they too are properties. The properties s, v and all functions thereof are called <u>internal properties</u> (thus u and b are called internal energy and internal availability respectively). The functions in Eqs. 27 through 30 are called <u>thermostatic constitutive relations</u> or <u>internal "equations" of state</u> -- or better, <u>internal functions of state</u>.

   Suppose, now, that starting from the dead state (where $s \equiv s_0$ and $v \equiv v_0$) the velocity $\underline{V}$ is ideally changed, while s is held constant -- i.e., while $\tilde{q} \equiv 0$ -- and while v is held constant. Then, the dependence of e and a upon $\underline{V}$ must be reflected by the invariants of $\underline{V}$ inasmuch as a and e are scalar quantities. The vector $\underline{V}$ has only one independent invariant; select $V^2 = \underline{V} \cdot \underline{V}$. Then

$$\rho \frac{De}{D\sigma} = \rho \frac{\partial e}{\partial(V^2)} \frac{D(V^2)}{D\sigma} \tag{31}$$

$$\rho \frac{Da}{D\sigma} = \rho \frac{\partial a}{\partial(V^2)} \frac{D(V^2)}{D\sigma} \tag{32}$$

A process of a simple compressible material is here said to be <u>non-relativistic</u> if $\partial a/\partial(V^2)$ and $\partial e/\partial(V^2)$ are independent of s,v, $\underline{V}$. Then symbolize these quantities by $1/2\,k_a$ and $1/2k_e$ respectively:

$$\frac{\partial a}{\partial(V^2)} \Bigg]_{s,v} \equiv \frac{1}{2k_a} \tag{33}$$

$$\left. \frac{\partial e}{\partial (v^2)} \right]_{s,v} \equiv \frac{1}{2k_e} \tag{34}$$

for non-relativistic simple compressible processes.   Then

$$e = u(s,v) + v^2/2k_e \tag{35}$$

$$a - p_0 v = b(s,v) + v^2/2k_a \tag{36}$$

Equations 35 and 36 are called <u>thermokinematic functions of state</u>. (Note that the variable s was introduced along with Eq. 23 in order to facilitate elimination of $\tilde{q}$ and $\tilde{r}$ from Eqs. 19 and 20 respectively. A more natural way to eliminate these variables would be to simply multiply Eq. 19 by α, then subtract from Eq. 20.   In the latter procedure entropy s would never be defined, rather a function for "internal availability" b(e,v) would arise.   The choice of introducing s was made in order that the traditional results would be obtained.)

The existence of the relations given by Eqs. 27, 28, 29 and 30 can be called the <u>internal state principle</u>.   The existence of the relations given by Eqs. 35, 36, 29 and 30 can be called the <u>non-relativistic thermokinematic state principle</u>.   (Note the role of α in the foregoing developments of the thermokinematic functions of state: By introducing α, covariant derivatives could be eliminated in favor of parameter derivatives of properties.   Thus, for example, Eq. 25 is obtained in lieu of Eq. 19.   Then by consideration of line integrals of the resultant equation from any given state to a common reference state, because the integrals are independent of the path, the functions of state can be deduced.   For example, Eq. 27 is deduced via Eq. 25.   Hence, α can be called an integrating factor.)

Next, the thermokinetic equations of change will be deduced. This is accomplished by deriving an expression for the availability destruction for real processes, and then invoking Postulate II, which decrees that a process will occur whenever it would destroy availability.   Consider that, differentiating Eq. 36 and using Eqs. 4 and 30 in the result yields

$$\rho \frac{Da}{D\sigma} = p_0 \nabla \cdot \underline{V} + \frac{\alpha}{1-\alpha} \rho \frac{Ds}{D\sigma} - p\nabla \cdot \underline{V} + \frac{\rho}{k_a} \underline{V} \cdot \frac{DV}{D\sigma} \tag{37}$$

Let $\underline{\beta}$ be the second order tensor such that

$$\underline{r} = \beta \cdot \underline{q} \tag{38}$$

Then, substituting Eqs. 22 and 38 into the difference between Eqs. 37 and 16, and solving for a_p gives

$$a_p = \underline{V} \cdot [\underline{\beta} \cdot \underline{q}] - \alpha \nabla \cdot \underline{q} - p[1-\alpha] \nabla \cdot \underline{V}$$
$$+ \rho[1-\alpha]\underline{V} \cdot \frac{D\underline{V}}{D\sigma} + [1-\alpha]\nabla \cdot [\underline{\pi} \cdot \underline{V}] \tag{39}$$

(By proper selection of scales (8) for availability, velocity and mass, k_a can be set equal to unity; for convenience, this has been done in Eq. (39) and will be done throughout the remainder of this paper.) Equation (39) can be rearranged into

$$a_p = \underline{q} \cdot [\nabla \cdot \underline{\beta}] + [\underline{\beta}^\dagger - \alpha \underline{g}] : \nabla \underline{q}$$
$$+ [1-\alpha] \; [\underline{\pi}^\dagger - p \underline{g}] : \nabla \underline{V} + [1-\alpha] \underline{V} \cdot \left[\rho \frac{D\underline{V}}{D\alpha} + \nabla \cdot \underline{\pi} \right] \tag{40}$$

Suppose that the material is opaque (3); i.e., that the local availability destruction at a particle is independent from that of the neighboring particles and depends only on the local internal state and the variables in Eq. 40. Also assume for the present that there is no coupling between the terms of Eq. 40. This assumption is made only for simplicity; it can be relaxed, but that complicates the following logic and conclusions.

Consider first, the special case such that, $\underline{V} \equiv \nabla \underline{q} \equiv 0$. Then, as will be discussed presently, it follows from Postulate II that

$$\rho \frac{D\underline{V}}{D\sigma} = - \nabla \cdot \pi \tag{41}$$

$$\underline{\beta} = \alpha \underline{g} \tag{42}$$

$$\underline{q} = L(s, v, \nabla \alpha) \tag{43}$$

$$\underline{\tau} \equiv \underline{\pi} - p \underline{g} = \Lambda(s, v, \nabla \underline{V}) \tag{44}$$

The foregoing are deduced from Eq. 40: If Eq. 41 were not satisfied, then it follows from Eq. 40 that availability would be destroyed were $\underline{V}$ non-zero; by Postulate II, $\underline{V}$ would then be non-zero. But this refutes the supposition that $\underline{V} = 0$, and hence Eq. 41 must be satisfied. Similar logic leads to Eq. 42. To deduce Eq. 43 note that since $\nabla \alpha = \nabla \cdot \underline{\beta} \neq 0$, on the basis of Eq. 40, Postulate II says that there will be $\underline{q} \neq 0$, when $\nabla \alpha \neq 0$, to destroy availability. Thus L yields zero when $\nabla \alpha = 0$. And, when $\nabla \alpha$ is non-zero, then $\underline{q}$ is non-zero and has the opposite direction to $\nabla \alpha$. Equation 44 is deduced analogously. The special case at hand is the case when V^2 is small -- in accord with the earlier restriction, viz. non-relativistic systems -- and when the invariants of $\nabla \underline{q}$ are small. Equations 41, 42, 43, and 44 are called the thermokinetic functions of state for this special class of problems at hand. Equation 41 is recognized as the traditional equation of change for momentum.

Now a complete set of governing equations is given, for example, by (a) the four thermokinetic functions of state above, along with (b) the thermokinematic functions of state -- Eqs. 27 and 28 --, (c) the local equations of change for mass, energy, and availability -- Eqs. 4, 12 and 16 --, and (d) the "auxiliary" relationships given by the symmetry of π, the definition of $\underline{\tau}$ (see Eq. 44), and Eqs. 5, 35, 36 and 38. (The equation of change for entropy -- equation 22 -- can be used in lieu of that for either availability or energy.)

It should also be mentioned that, for L and Λ isotropic functions, truncated Maclaurin expansions of Eqs. 43 and 44 and use of Eq. 29 to change the independent variables from (s,v) to (α,p) results in the traditional forms for Fourier's law of heat condition and Newton's law of viscosity, viz.

$$\underline{q} = - \kappa(p,\alpha)\nabla\alpha \tag{45}$$

and

$$\underline{\tau} = \lambda(p,\alpha) \ [\nabla\cdot\underline{V}]\underline{g} - \mu(p,\alpha) \ [\nabla\underline{V} + \nabla\underline{V}^{\dagger}] \tag{46}$$

(It is readily shown that α defines a Temperature scale.) The theory has thus derived these "laws" as truncated expansions of a Maclaurin series, and shows that transport properties are dependent upon (α,p). The theory also predicts the existence of (α,p) dependence for u,b, and v. However, the functions themselves must be determined via experiment. The methods for establishing these functions from experimentally measurable functions are not a part of this paper, inasmuch as they follow the lines presented in the typical presentations of thermodynamics.

Governing Equations for a Simple Compressible Reactive Multiconstituent Flow (with Internal Constraints)

The case to be considered here is that of a continuum mixture of chemical species in which chemical reactions occur. (A lumped-parameter model is given in (2).) The model for each different chemical compound is called a constituent. Were the constituents inert, then the amount of each would be an external constraint. However, since compounds can be produced or destroyed by chemical reactions in the mixtures being considered, the amount of each constituent is a non-conserved property. Then, if a mixture contains N constituents, and if R independent reactions are allowed to take place, N - R of the constituents can be selected as components; and the amount of each component is a conserved property, i.e. its value can change only by being transported -- it cannot be produced. Thus, every component is a constituent, but not every constituent is a component. However, the component amount, and the constituent amount of the same species are different. The amount of a constituent in a mixture at any instant represents the

actual amount present. On the other hand, given a set of components, the amount of each component does not represent the amount
present (except in inert mixtures), rather it represents the
amount of that component that would be required to produce the
mixture from that component set. A more elaborate discussion on
the distinction between constituents and components is given in
($\underline{3}$, pp. 193–194).

The number of components C in a mixture equals the number of
constituents N minus the number of independent reactions R. It is
always possible to select R independent reactions which are so-
called formation reactions: A formation reaction is one which
yields only one constituent, from all the components. For the formation-reaction of constituent β, the stoichiometric coefficient
of β is unity; the stoichiometric coefficient of component α is
denoted by $\{-\nu_{\beta\alpha}\}$, $\alpha = 1,\ldots, C$.

The amount per unit volume of component α at place p at "time"
σ will be designated $\gamma_\alpha(p;\sigma)$ and the amount per unit volume of
constituent β at place p at "time" σ will be designated ρ_β. The
sum

$$\rho(p;\sigma) \equiv \sum_{\alpha=1}^{C} \gamma_\alpha(p;\sigma) = \sum_{\beta=1}^{N} \rho_\beta(p;\sigma) \tag{47}$$

then represents the total amount of "material" per unit volume at
place p at "time" σ. (For simplicity of the presentation to follow, these amounts are expressed in "inertial" units (e.g. grams,
pounds mass) rather than "mole" units (e.g. gram moles, pound
moles).) Hence the set $(\rho,\gamma_1, \ldots \gamma_C)$ is not an independent set
of properties; any one of these can be found in terms of the other
C.

In addition, y_α and x_β will designate the fraction amounts of
component α and constituent β respectively; that is

$$y_\alpha(p;\sigma) \equiv \frac{\gamma_\alpha(p;\sigma)}{\rho(p;\sigma)} \qquad \alpha = 1, \ldots C \tag{48}$$

$$x_\beta(p;\sigma) \equiv \frac{\rho_\beta(p;\sigma)}{\rho(p;\sigma)} \qquad \beta = 1, \ldots N \tag{49}$$

It follows that

$$\sum_{\alpha=1}^{C} y_\alpha = \frac{1}{\rho}\sum_{\alpha=1}^{C} \gamma_\alpha = \sum_{\beta=1}^{N} x_\beta = \frac{1}{\rho}\sum_{\beta=1}^{N} \rho_\beta = 1 \tag{50}$$

Thus the set $(\rho,y_1, \ldots y_{C-1})$ or the set $(\gamma_1, \ldots, \gamma_C)$.
is equivalent to any independent set from $(\rho,\gamma_1,\ldots \gamma_C)$.

The fraction amount x_β of each constituent β which is not a component, is called the formation reaction extent for that species; it represents the extent to which the formation reaction for that species has proceeded. The amount of component α which has been "used up" by reaction β to form constituent β is given by $- \nu_{\beta\alpha} x_\beta$. It then follows that the fraction amount of component α is related to the fraction amount of constituent α by

$$ y_\alpha = x_\alpha - \sum_{\beta=C+1}^{N} \nu_{\beta\alpha} x_\beta \qquad \alpha = 1, \ldots C \qquad (51) $$

where species "C+1 through N" are taken to be those constituents which are not components, and species "1 through C" are taken to be the components. This notational convention will be used throughout the rest of this section.

An equilibrium state is one which would not change were it isolated. Thus, considering the intensive states of the model at hand, if the specific energy, velocity, specific volume, and the amounts of each component in a set are uniquely specified, then no interaction with the environment would be possible. Hence, if consideration is restricted to simple compressible equilibrium states, they are fixed by specification of $(e, \underline{V}, v, y_1, \ldots, y_{C-1})$. Hence, considering only equilibrium states, the amount of each constituent is fixed by specification of $(e, \underline{V}, v, y_1, \ldots, y_{C-1})$.

If it is desired to consider also states which are not at reactive equilibrium -- which could reach the equilibrium states via chemical reaction at fixed $(e, \underline{V}, v, y_1, \ldots, y_{C-1})$ -- then specification of these variables clearly does not suffice to fix the constitution. Although, by definition, the states being considered are defined by $(e, \underline{V}, v, x_1, \ldots, x_{N-1})$, it will be more convenient to describe them with $(e, \underline{V}, v, y_1, \ldots, y_{C-1})$ plus the formation-reaction extents, i.e. with $(e, \underline{V}, v, y_1, \ldots, y_{C-1}, x_{C+1}, \ldots, x_N)$.

In this section the "total amount of components" fulfills the same role as "mass" in the preceding section. Hence Eq. 4 is now regarded as the balance equation for the total amount of components. For compatibility with this, the velocity $\underline{V}$ is now defined as

$$ \underline{V} \equiv \frac{1}{\rho} \sum_{\alpha=1}^{N} \rho_\alpha \underline{V}_\alpha \qquad (52) $$

and $\underline{V}_\alpha$ is the velocity of the αth constituent manifold.

An equation of change for the formation reaction extent x_β is

$$ \rho \frac{Dx_\beta}{D\sigma} = - \nabla \cdot \underline{j}_\beta + \rho r_\beta \qquad \beta = C + 1, \ldots, N \qquad (53) $$

where

$$\underline{j}_\beta \equiv \rho_\beta [\underline{V}_\beta - \underline{V}] \qquad \beta = 1, \ldots, N \qquad (54)$$

Note that Eqs. 52 and 54 yield

$$\sum_{\alpha=1}^{N} \underline{j}_\alpha = 0 \qquad (55)$$

Thus $\underline{j}_\beta$ represents the flux of constituent β relative to the bulk velocity of the mixture (i.e. the diffusion flux -- and this has been defined for all constituents), and r_β represents the net production rate of constituent β per unit amount of mixture. The presence of this production term reflects the fact that the constituent amounts are not conserved.

If constituent α is also a component, then there is a possibility that the species can be destroyed by more than one reaction. Hence the equations of change for x_α, $\alpha = 1, \ldots C$, will appear as

$$\rho \frac{Dx_\alpha}{D\sigma} = - \nabla \cdot \underline{j}_\alpha + \rho \sum_{\beta=C+1}^{N} \nu_{\beta\alpha} r_\beta \qquad \alpha = 1, \ldots C \qquad (56)$$

The term $\nu_{\beta\alpha} r_\beta$ represents the net production rate of constituent α by formation reaction β per unit amount of mixture.

An equation of change for the fraction component amount y_α is

$$\rho \frac{Dy_\alpha}{D\sigma} = - \nabla \cdot \underline{j}_\alpha + \sum_{\beta=C+1}^{N} \nu_{\beta\alpha} \nabla \cdot \underline{j}_\beta \qquad \alpha = 1, \ldots C \qquad (57)$$

Note the absence of production terms, which is to be expected since component amounts are conserved.

Next, balance equations for energy and availability are written as

$$\rho \frac{De}{D\sigma} = - \nabla \cdot \left[\underline{\pi} \cdot \underline{V} + \sum_{\substack{\alpha=1 \\ \alpha \neq C}}^{N} \delta_\alpha \underline{j}_\alpha + \underline{q} \right] \qquad (58)$$

and

$$\rho \, \frac{Da}{D\sigma} = - \, \nabla \cdot \left[[\underline{\pi} - p_0 \underline{g}] \cdot \underline{V} + \sum_{\substack{\alpha=1 \\ \alpha \neq C}}^{N} [\delta_\alpha - \delta_{0\alpha}] \underline{j}_\alpha + \underline{r} \right] + a_p \qquad (59)$$

In writing these the volume plus the amount of each consti-
tuent except the Cth, i.e. $\{v, x_1, \ldots x_{C-1}, x_{C+1}, \ldots x_N\}$, were
taken as an independent set of constraints for the system. (Alternatively, one could take the amount of each of the N constituents
as the independent constraints. In one case, the N − 1 constituent manifolds and the overall mass manifold are selected as independent; in the other case, the N constituent manifolds are selected.) A term reflecting the flux of energy and availability
with the flux of each of these constraints was incorporated in
each of the respective balance equations. That such terms exist
follows from Postulate I and II. A further assumption underlies
these balance equations, namely that only the volume transport,
represented by $\rho \underline{V}$, has inertia, and that the effects of this inertia are reflected by the same momentum equation as that in the
preceding section; viz. Eq. 41. Were there inertia associated
with, for example, the flux j_α, then the scalar coefficient δ_α
would have to be replaced by a second order tensor. Consistent
with this assumption is the further assumption that during ideal
processes

$$\nabla \cdot [\tilde{\delta}_\alpha \underline{j}_\alpha] = \tilde{\delta}_\alpha \nabla \cdot \underline{j}_\alpha \Bigg]_{ideal} \qquad (60)$$

The next step in order to derive the pertinent thermokinematic functions of state, is to write Eqs. 58 and 59 for ideal
processes, and then combine the results to eliminate covariant derivatives of fluxes in favor of coefficients times parameter derivatives of specific intensive properties. In the previous section
this could be accomplished merely by multiplying one equation by
α and then subtracting the other. Before that can be done here,
however, the terms in the summation signs must be expressed in
terms of parameter derivatives. This is done by using Eqs. 56 and
57. The results are

$$\rho \, \frac{Du}{D\sigma} \Bigg]_{ideal} = - \, p\rho \, \frac{Dv}{D\sigma} + \rho \sum_{\alpha=1}^{C-1} \tilde{\delta}_\alpha \, \frac{Dy_\alpha}{D\sigma} + $$

$$+ \rho \sum_{\beta=C+1}^{N} \tilde{\lambda}_\beta \left[\frac{Dx_\beta}{D\sigma} - r_\beta \right] \quad - \nabla \cdot \tilde{\underline{q}} \tag{61}$$

and

$$\rho \frac{Db}{D\sigma} \bigg]_{ideal} = - p\rho \frac{Dv}{D\sigma} + \rho \sum_{\alpha=1}^{C-1} [\tilde{\delta} - \tilde{\delta}_{0\alpha}] \frac{Dy_\alpha}{D\sigma}$$

$$+ \rho \sum_{\beta=C+1}^{N} [\tilde{\lambda}_\beta - \tilde{\lambda}_{0\beta}] \left[\frac{Dx_\beta}{D\sigma} - r_\beta \right] \quad - \nabla \cdot \tilde{\underline{r}} \tag{62}$$

where

$$\lambda_\beta \equiv \delta_\beta + \sum_{\alpha=1}^{C-1} \delta_\alpha \nu_{\beta\alpha} \quad \beta = C=1, \ldots N \tag{63}$$

and u and b are defined as in the previous section (Eqs. 35 and 36. (It is readily shown that $\tilde{\lambda}_\beta$ is the chemical affinity of reaction β.)

It still remains to express the R reaction rates in terms of property derivatives. This is done by again employing a scheme analogous to Eq. 23, that is, define

$$\varepsilon_\beta \equiv r_\beta \bigg/ \frac{Dx_\beta}{D\sigma} \bigg]_{ideal} \quad \beta = C + 1, \ldots N \tag{64}$$

Thus ε_β is seen to represent the rate at which constituent β is being "formed" by reaction to the total rate of change of that constituent. Furthermore, if the mixture is inert, then $\varepsilon_\beta = 0$ for all $\beta = C+1, \ldots N$, and each formation reaction extent (each x_β, $\beta = C+1, \ldots N$) becomes a conserved property; therefore the ideal process is reversible (all the constraints are external -- i.e. conserved).

Substituting Eq. 64 in turn into Eqs. 61 and 62 yields

$$\rho \left.\frac{Du}{D\sigma}\right]_{ideal} = p\rho \frac{Dv}{D\sigma} + \rho \sum_{\alpha=1}^{C-1} \tilde{\delta}_\alpha \frac{Dy_\alpha}{D\sigma}$$

$$+ \rho \sum_{\beta=C+1}^{N} \tilde{\lambda}_\beta [1-\varepsilon_\beta] \frac{Dx_\beta}{D\sigma} - \nabla \cdot \tilde{q} \tag{65}$$

and, (using that $\tilde{\delta}_{0\alpha}$ is a constant for a particle)

$$\rho \left.\frac{D\left[b+ \sum_{\alpha=1}^{C-1} \tilde{\delta}_{0\alpha}y_\alpha\right]}{D\sigma}\right]_{ideal} = - p\rho \frac{Dv}{D\sigma} + \rho \sum_{\alpha=1}^{C-1} \tilde{\delta}_\alpha \frac{Dy_\alpha}{D\sigma}$$

$$+ \rho \sum_{\beta=C+1}^{N} [\tilde{\lambda}_\beta - \tilde{\lambda}_{0\beta}][1-\varepsilon_\beta] \frac{Dx_\beta}{D\sigma} - \nabla \cdot \tilde{r} \tag{66}$$

Now Eq. 23 can be employed to eliminate the remaining non-parametric derivatives, viz. $\nabla \cdot \tilde{q}$ and $\nabla \cdot \tilde{r}$. Multiplying Eq. 65 by α and subtracting the result from Eq. 66 gives, after some re-arranging:

$$\frac{D\left[b+ \sum_{\alpha=1}^{C-1} \tilde{\delta}_{0\alpha}y_\alpha + \sum_{\beta=C+1}^{N} \tilde{\lambda}_{0\beta}x_\beta\right]}{D\sigma}$$

$$= \alpha\frac{Du}{D\sigma} - [1-\alpha]p \frac{Dv}{D\sigma} + [1-\alpha] \sum_{\alpha=1}^{C-1} \tilde{\delta}_\alpha \frac{Dy_\alpha}{D\sigma}$$

$$+ \sum_{\beta=C+1}^{N} \left[[1-\alpha]\tilde{\lambda}_\beta[1-\varepsilon_\beta] + \varepsilon_\beta\tilde{\lambda}_{0\beta}\right]\frac{Dx_\beta}{D\sigma} \tag{67}$$

Now, as before, it follows that there exists a function f, such that

$$b + \sum_{\alpha=1}^{C-1} \tilde{\delta}_{0\alpha} y_\alpha + \sum_{\beta=C+1}^{N} \tilde{\lambda}_{0\beta} x_\beta$$

$$= f(u,v,y_1, \cdots y_{C-1}, x_{C+1}, \cdots x_N) \qquad (68)$$

and

$$\alpha = f_1(u,v, \cdots) \qquad (69)$$

$$-[1 - \alpha]p = f_2(u,v, \cdots) \qquad (70)$$

$$[1 - \alpha]\tilde{\delta}_\alpha = f_{\alpha+2}(u,v, \cdots) \quad \alpha = 1, \cdots, C-1 \qquad (71)$$

$$[1-\alpha]\tilde{\lambda}_\beta [1-\varepsilon_\beta] + \varepsilon_\beta \tilde{\lambda}_{0\beta} = f_{\beta+1}(u,v, \cdots) \quad \beta = C+1, \cdots N \qquad (72)$$

By procedures similar to those used in obtaining Eq. 40, the rate of availability destruction per unit volume is found to be

$$- a_p = - [1 - \alpha] \left[[\underline{\pi} - p\underline{g}]: \nabla \underline{V} + \sum_{\substack{\alpha=1 \\ \alpha \neq C}}^{N} \underline{j}_\alpha \cdot \nabla \delta_\alpha \right.$$

$$+ \rho \sum_{\beta=C+1}^{N} \left[r_\beta - \varepsilon_\beta \frac{Dx_\beta}{D\sigma} \right] \left[\tilde{\lambda}_\beta - \frac{\lambda_{0\beta}}{1-\alpha} \right]$$

$$+ \sum_{\substack{\alpha=1 \\ \alpha \neq C}}^{N} [\delta_\alpha - \tilde{\delta}_\alpha] \nabla \cdot \underline{j}_\alpha + \frac{q}{1-\alpha} \cdot \nabla \alpha \right] \qquad (73)$$

From this the following thermokinetic functions of state are deduced:

$$\underline{q} = \underline{q}(u,v,y,x,\nabla\alpha,\nabla\delta) \qquad (74)$$

$$\underline{j}_\alpha = \underline{j}_\alpha(u,v,y,x,\nabla\delta,\nabla\alpha) \quad \alpha = 1, \cdots C-1, C+1, \cdots N \qquad (75)$$

$$\underline{\pi} - p\underline{g} = \underline{\tau}(u,v,y,x,[\nabla\underline{V} + \nabla\underline{V}^\dagger], \tilde{\lambda}, \nabla\cdot\underline{j}) \qquad (76)$$

$$\delta_\alpha - \tilde{\delta}_\alpha = \phi_\alpha(u,v,y,x,\nabla\cdot\underline{j},\tilde{\lambda},\nabla\cdot\underline{V}) \quad \alpha = 1, \cdots C-1, C+1, \cdots N \qquad (77)$$

$$r_\beta - \varepsilon_\beta \frac{Dx_\beta}{D\sigma} = R_\beta(u,v,y,x,\tilde{\lambda},\nabla\cdot\underline{j},\nabla\cdot\underline{V}), \quad \beta = C+1,\ldots N \qquad (78)$$

$$\lambda_{0\beta} = 0 \qquad \beta = C+1,\ldots N \qquad (79)$$

(In the lists of independent variables, the subscripts have been
deleted from x,y,δ,$\underline{j}$ etc. for convenience.) The deduction of
these equations from the above expression for a_p, is analogous to
that employed previously. Note that Eq. 79 follows from the fact
that the intensive state of the complete stable equilibrium
configuration of a particle is constant and that $\delta_{0\alpha}$ represents
the amount of energy transported.

In summary then, a complete set of governing equations is
given by:

  Equations of change: Eqs. 4, 56, 57, 58, 59 and 41.
  Thermokinematic functions of state: Eqs. 68 through 72.
  Thermokinetic functions of state: Eqs. 74 through 78.
  Auxiliary equations: Eqs. 5, 35, 36, 38, 50 and 55.
These constitute (22 + 7R + 6C) equations in the (22 + 7R + 6C)
unknowns p, v, e, u, a, b, α, p, a_p, $\underline{V}$(3), j_α(3R+3C), r_β(R),$\tilde{\delta}_\alpha$
(C-1), $\underline{q}$(3), $\underline{\beta}$(3), λ_β(R), ε_β(R), δ_α(C-1), y_α(C), x_β(R). Also, if
the six equations for Da/Dσ, f(u,v,y,x), $\underline{r} = \underline{\beta}\cdot\underline{q}$, a = b + p_0v +
$V^2/2k$ are deleted then the remaining system could be solved for
all the variables except a,b,a_p, $\underline{r}$. It should also be mentioned
that other alternative sets of independent governing equations
could be selected, via changes of variables. For example, simple
rearrangement of Eq. 67 would yield a function for s = b - u
rather than for b.

With such alternatives, different but theoretically equivalent sets of governing equations could be deduced; which would
prove to be the more easily and more economically employed sets
would be largely a matter of experience and insight. In any case,
the procedure outlined in this paper resolves the basic difficulty:
finding a set -- any set -- of pertinent governing equations.
(Note, though, that the procedure does not yield the functions of
state, but only tells what their variables will be. But simply
knowing what they will be greatly simplifies the determination,
via experimentation, of the functions themselves.)

The most notable difference between the equations developed
here and the standard reaction equations, is the appearance, for
each formation reaction, of the variable ε_β in the governing equations. The new variable enters as an integrating factor which is
needed because the states are far from equilibrium. This is in
contrast to traditional theories (e.g., 6, Chapter X), wherein it
is assumed that all states, regardless of composition, can be connected reversibly. The traditional theories have proven valuable—
i.e., the assumption "works out" — for predicting equilibrium
compositions; see (2, Appendix B.5). But at best, "empirical modifications" of these traditional theories have been required to

obtain good results for <u>reacting</u> mixtures. This failure of the
traditional theories can be explained from the results obtained
in this paper, in that the traditional theories regard the affinity -- an "equilibrium" variable -- as the driving force for reaction, whereas the results here (viz. Eq. 78) show that ε_β must
also be included. Thus the theory given here includes not only
the traditional states of equilibrium composition, but also the
highly nonequilibrium states which are encountered in "real" chemical reactions.

Closure

It has been illustrated how, by replacing the reversibility
concept by the more general concept, ideality, a thermokinematic
and a thermokinetic state principle can be derived which is applicable to nonequilibrium states. The key to replacing reversibility
by ideality is to base the second law on availability; the key to
implementing this for the purpose of deducing property relations
is the introduction of an integrating factor, in analogy to temperature. These state principles along with the extensive property balance equations provide a system of governing equations.

Abstract

    In thermodynamics, entropy has been defined only for equili-
brium states. In the present approach this is overcome by replacing the concept of reversibility in favor of a more general concept,
called ideality, which includes reversibility as a special case.
    There are two keys to this generalization, whereby a rational
definition for entropy of nonequilibrium states is obtained. First,
a more general second law is postulated, based on the property
<u>availability</u>. (The availability at any state of a system reflects
the extent to which it could affect any other system.) The second
key is the introduction of other integrating factors, in addition
to temperature, in order to deduce the fundamental differential
property relationship (i.e., Gibbs equation.)
    Another feature of the present theory is that it provides a
formalism for deducing a complete mathematical representation of a
phenomenon. In particular, beginning with the balance equations
for (1) the pertinent constraints (e.g. volume, alone, for "simple
compressible flows"), (2) the energy and (3) the availability, the
needed complementary equations are deduced; namely, the equations
reflecting inertial effects (e.g., the momentum equation), and the
constitutive relations. In this paper, the theory is applied to
two specific cases. The first case is simple compressible flows
to illustrate the theory in familiar circumstances. (Where some
assumptions inherent in the traditional governing equations also
come to light.) The second case is chemically reactive flows;
here the added generality of ideality over reversibility is needed, since the states visited are not at internal equilibrium. As

desired, the governing equations obtained distinguish themselves from those currently employed, by the appearance of relations involving additional integrating factors, besides temperature.

Other highly nonequilibrium phenomena for which the added generality of the proposed theory could be fruitful, inasmuch as the contemporary models are not satisfactory, are multiphase metastable flows and viscoelastic flows.

Acknowledgment

The authors would like to acknowledge the counsel of Professor Edward F. Obert. Those who are familiar with (3) will recognize its influence, which was crucial.

Literature Cited

1. Obert, E.F. and Gaggioli, R.A. Thermodynamics. McGraw-Hill, New York, Second Edition, 1963.
2. Gaggioli, R.A. and Scholten, W.B. "A Thermodynamic Theory for Non-equilibrium Processes I. Thermokinematics." Marquette University College of Engineering Report, Marquette University, Milwaukee, Wisconsin, Nov., 1970.
3. Hatsopoulos, G.N. and Keenan, J.H. Principles of General Thermodynamics, John Wiley & Sons. New York, 1965.
4. Sokolnikoff, I.S. Tensor Analysis Theory and Applications to Geometry and Mechanics of Continua. John Wiley & Sons, New York, Second Edition, 1964.
5. Lodge, A.S. Elastic Liquids. Academic Press, New York, 1964.
6. deGroot, S.R. and Mazur, P. Non-Equilibrium Thermodynamics. North-Holland, Amsterdam, 1962.
7. Wepfer, W.J. "Applications of the Second Law to the Analysis and Design of Energy Systems." Ph.D Dissertation, University of Wisconsin-Madison, 1979.
8. Moran, M. and Gaggioli, R. "Generalized Dimensional Analysis." U.S Army Mathematics Research Center Report No. 927, 1968.

RECEIVED October 17, 1979.

Nonequilibrium Thermodynamics

LUC LEPLAE

Department of Physics, University of Wisconsin–Milwaukee,
Milwaukee, WI 53201

Steady states are among the phenomena that non-equilibrium
thermodynamics studies. When one is not too far from equilibrium
it can be shown that the steady states are stable. On the other
hand, when far from equilibrium, certain systems can make transitions to states exhibiting "dissipative structures." The theory
of non-equilibrium developed by I. Prigogine, is quite general
and has been applied to a wide range of phenomena. It is the
aim of this lecture to introduce this field with a few examples.

Introduction

Self Organization

When the temperature of a ferromagnet is brought below the
Curie temperature a new order takes place: the atomic magnetic
moments, which above T_c were oriented at random, align themselves
with each other under T_c, forming magnetic domains.

The creation or destruction of order in phase transitions is
a well known and well studied fact. What is not so well known
is that self organization also takes place in an entirely different class of phenomena. Certain systems when far away from
equilibrium jump to new states where new structures appear.
These structures have been called dissipative structures by
Ilya Prigogine, who gave the first general thermodynamic description of these phenomena ([1],[2]). He has been awarded the 1977
Nobel Prize in Chemistry for that work.

0–8412–0541–8/80/47–122–227$05.00/0
© 1980 American Chemical Society

B. Examples of Non Equilibrium Systems

To understand better what will follow it may be helpful to have in mind a few concrete and simple examples. For instance:

(i) a system consisting of two large heat reservoirs A and B connected by a small heat conducting system C (Figure 1a).

(ii) a system consisting of two large water tanks connected by a thin pipe (Figure 1b).

(iii) two electrically charged spheres connected by an electric conductor (Figure 1c).

In each case we are interested in the behavior of C.

When $T_A > T_B$ heat flows through C, when $h_A > h_B$ water flows through C, when $V_A > V_B$ an electric current flows through C.

The Linear Region

Onsager Reciprocity Relations (3)

When one is still very close to equilibrium ($T_A \overset{\sim}{>} T_B$, $h_A \overset{\sim}{>} h_B$, $V_A \overset{\sim}{>} V_B$ in our three examples) a flow appears in C. To these flows are also associated some generalized forces. For instance in the electric example, an electric field $\vec{E} = \vec{\nabla}V$ appears along the wire which induces the electric current. In the thermal example, a temperature gradient is created along the system C which induces the heat flow.

At equilibrium both flows and generalized forces vanish. When very close to equilibrium the flows can be written as linear expressions in the generalized forces. For instance, $\vec{J} = \sigma\vec{E}$ in the electric case, which is nothing but ohm's law, or $\frac{dQ}{dt} = \kappa \frac{A}{L} \nabla T$ in the thermal case, which is the heat transfer equation.

In general there can be several types of flows J_k through a given system C, and several generalized forces X_i producing them (for instance, a conducting rod through which flows an electric and a heat current produced by an electric field and a temperature gradient).

Close to equilibrium one can thus write the general relation

(1) $J_k = \sum_i L_{ki} X_i$

The L_{ki} are called the phenomenological constants. This region
is called the linear region.

Onsager showed in 1931 that with proper choice of fluxes and
forces, $L_{ki} = L_{ik}$. These relations are now called the Onsager
reciprocity relations.

Stability of the Steady State in the Linear Region

If in our three examples A and B are very large compared to
C, the flow through C does not modify A and B. If the flow
through C is constant, everything is essentially time independent, including the boundary conditions of the system C. In that
case one says that C is in a steady state although it is not in
thermodynamic equilibrium.

The systems C are open systems: they can exchange energy
and matter with the exterior (systems A and B).

If we assume that the systems C obey local equilibrium
thermodynamics (see Ref. 2, page 30) an entropy can be associated
to these non equilibrium systems and one can show that the total
change of their entropy can be written as

(2) $dS = d_iS + d_eS$

where d_eS is the change due to a flow of entropy from A and B,
whereas d_iS is a change of entropy produced by irreversible
processes taking place in C.

We will call $P \equiv \dfrac{d_iS}{dt}$ the entropy production rate. One can
show that

(3) $P = \int dv \, [\sum_i X_i \, J_i]$

where X_i, J_i are the flows and generalized forces present in the
system C, and the integration is a volume integral over C.

Prigogine has shown that if the boundary conditions of the

systems C are kept constant and if one is in the linear region, then

(4) $\quad \dfrac{dP}{dt} \lessgtr 0$

As on the other side $P \equiv \dfrac{d_i S}{dt} \gtrless 0$ (irreversible processes can only increase the entropy), one concludes that the system C will evolve spontaneously to a state where $\dfrac{dP}{dt} = 0$, where the entropy production rate is minimum, this state being a steady state.

The behavior of P is illustrated in Figure 2.

In other words if the system C is not too far away from equilibrium, and its boundary conditions are kept constant, it will eventually reach a steady state. This steady state is stable, and corresponds to a minimum entropy production rate. This is called the minimum entropy production theorem (1,2).

We can illustrate the theorem using our thermal example. The graphs (a), (b) and (c) show the evolution of the temperature of C (Figure 3). In graph (c), which corresponds to the steady state, we have assumed for simplicity that the coefficient of thermal conduction is independent of temperature.

The Non Linear Region

Stability Conditions (2)

When a system is far away from equilibrium and outside the linear region, one cannot say anything about the sign of $\dfrac{dP}{dt}$: the steady state is not necessarily stable. In fact, in many cases it is unstable and the system can jump to new states, the dissipative structure states. If we want to determine whether the steady state is stable, we can perturb the system slightly away from the steady state and study its behavior. Let $\Delta S \equiv S - S_o \cong \delta S + \frac{1}{2} \delta^2 S$ be the change of entropy due to this perturbation, S_o being the entropy at the steady state. S is a functional of the local thermodynamic variables. In δS we have collected all the terms linear in the change of these variables

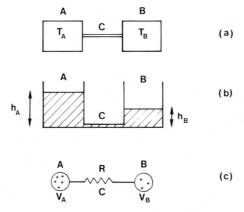

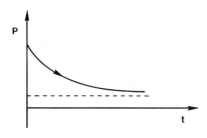

Figure 1. Examples of nonequilibrium systems: (a) thermal system; (b) hydro-dynamic system; (c) electric system.

Figure 2.   Time evolution of the entropy production rate in the linear region

and in $\delta^2 S$ all the terms quadratic in those changes.

One can show that if one does not go too far from the steady state:

$$\delta S = 0$$

(5) $\delta^2 S \leq 0$

(6) $\dfrac{d}{dt} \delta^2 S = \displaystyle\int dv \; [\sum_k \delta X_k \; \delta J_k] \equiv 2\delta_x P$

$\delta_x P$ is called the <u>excess</u> entropy production rate. If $\delta_x P \equiv \dfrac{1}{2} \dfrac{d}{dt} \delta^2 S > 0$, ΔS will eventually go to zero: a perturbation of the steady state goes to zero with time, and the steady state is stable. If on the other side $\delta_x P < 0$, a small perturbation is amplified and the steady state is unstable. These inequalities are called the stability conditions. (See Figure 4.)

Bénard Cells as an Example of Dissipative Structures

A well known example of dissipative structures are the Bénard cells, also called convection cells. A flat tank is filled with water. The upper and lower surfaces are kept at different temperatures. When the lower surface is slightly warmer than the upper one, the system is in the linear region and one observes a steady and uniform upward flow of heat.

If the temperature of the lower surface is increased, suddenly convection currents appear and very regular structures are formed which when looked at from above form a hexagonal pattern. These structures have been observed by Bénard for the first time in 1901, and are very well known by meteorologists.

Systems Involving Chemical Reactions (2)

The Basic Equations

Up to now the argument has been very general. We saw that some systems when far from equilibrium jump from the steady state to dissipative states, and we saw that the stability depends on

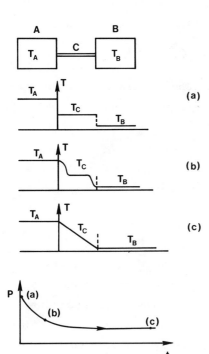

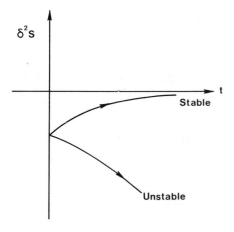

Figure 3.   *Time evolution of a nonequilibrium system towards a steady state in the linear region*

Figure 4.   *The time evolution of the excess entropy production $\delta^2 S$ in the nonlinear region gives a criterion for the stability of the steady state*

the sign of $\delta_x P$, the excess entropy production rate.

If we want to learn more about those dissipative structures, we have to study precise examples. The type of systems that Prigogine and his school have decided to study in detail are composed of a medium (a liquid solvent or an inert gas) in which are diluted n chemical components. The temperature of the system is assumed to be constant, and the system is assumed to be at mechanical equilibrium (no mass flow) and is not subject to external fields.

If we call ρ_i the density of the ith chemical component, one can show that the following equation is true:

$$(7) \qquad \frac{\partial \rho_i}{\partial t} = D_i \nabla^2 \rho_i + F_i(\{\rho_k\})$$

The D_i are diffusion coefficients and the F_i are non linear functions of the ρ_k. This last term comes from the reaction rates. These equations have been called the basic equations.

As one can see we are dealing with non linear differential equations. There exists no general method for solving this type of equation. In order to be able to solve them one has to simplify the model as much as possible.

The Tri-molecular Model

This model consists of: (1) two initial chemical components A and B which are absorbed by the system from outside. Their densities are kept constant throughout the system by a continuous supply from outside. (2) Two final chemical components D and E which are rejected by the system. Their densities are kept constant artificially too, by extracting them continuously. (3) Two intermediate components X and Y, the densities of which are variable.

Prigogine and co-workers have chosen the simplest reactions among those components for which instability can take place

(8) $A \overset{\rightarrow}{\underset{\leftarrow}{}} X$

         $B + X \overset{\rightarrow}{\underset{\leftarrow}{}} Y + D$

         $2X + Y \overset{\rightarrow}{\underset{\leftarrow}{}} 3X$

         $X \overset{\rightarrow}{\underset{\leftarrow}{}} E$

and found the following system of equations.

(9)
$$\begin{cases} \dfrac{\partial X}{\partial t} = A - (B+1)X + X^2Y + D_1 \nabla^2 X \\[2mm] \dfrac{\partial Y}{\partial t} = Bx - X^2Y + D_2 \nabla^2 Y \end{cases}$$

where X, A, etc. are proportional to the densities of the corresponding components, and where D_1 and D_2 are the diffusion coefficients of components X and Y, respectively.

Under this form the equations are still too hard to solve. One has to make a final simplification, and assume that the system is essentially a one dimensional system of length ℓ.

Depending on the boundary conditions (i.e. the values of D_1, D_2 and A), the dissipative structures are time independent or time dependent. For the time independent case one finds the following results.

(10) $X = X_0 \pm \left(\dfrac{B-B_c}{\phi}\right)^{\frac{1}{2}} \sin \dfrac{m_c \pi r}{\ell} + \ldots$

where X_0 corresponds to the steady state. These solutions can be represented graphically (Figure 5).

The density of A being kept constant, the system is brought away from equilibrium by varying B. B_c is the critical value at which the system bifurcates from the steady branch (a) to one of the dissipative structure branches (b) or (c).

The expressions for X, Y show that these concentrations vary in space as $\sin \dfrac{m_c \pi r}{\ell}$ where m_c is an integer which depends on the values of D_1, D_2 and A.

If we imagine our one dimensional system as a long test tube in which the reactions take place, the branch (a) of Figure 5 would correspond to the case where the concentrations X and Y are constant along the tube, whereas branch (b) or (c) would corres- pond to the case where the concentrations X and Y vary periodi- cally along the tube forming an array of layers which can be observed. Illustrations corresponding to this type of experi- ments are given in Refs. 1 and 2.

The Belousov-Zhabotinski Reactions

Experimentally dissipative structures in chemical systems have been observed. The best known are probably the Belousov- Zhabotinski reactions. Already two articles describing them have appeared in Scientific American, in the issues of June 1974 and July 1978.

The Josephson Effects

Introduction

We have seen in the general introduction that self ordering can occur in two very different situations: at equilibrium, when a system goes through a phase transition (example of ferromagne- tism) or far away from equilibrium when dissipative structures appear in the system (example of Bénard cells).

We will see in this section that there exists a third type of phenomena, the Josephson effects, which occupy a position between phase transitions and dissipative structures.

The analogy between phase transition and bifurcations to dissipative structures has been noted by several authors. Some authors (4) are even looking for a unified formalism that could describe both of these phenomena. The study of the Josephson effects can thus be a useful tool in this line of thought.

We will describe two different types of Josephson junctions:
(i) The Anderson-Dayem bridges, the behavior of which can be
    interpreted very nicely in terms of dissipative structures,
    but for which the theoretical description is not so simple.

(ii) The oxide barrier Josephson junctions, which can be des-
      cribed by rather simple theoretical models.

The Anderson-Dayem Bridges (5)

An Anderson-Dayem bridge consists of a thin superconducting
film having the shape represented in Figure 6. The narrowed
part of the bridge, NP, is typically of the order 3μ. The potential difference V_o is progressively increased starting from zero.
The current going through the bridge and the potential drop
across the bridge are measured. (Figure 7)

At first a current is observed but no potential drop. The
current is a supercurrent, and the system is said to be in the
pure superconducting state (P.S. state). This situation continues until a critical current I_c is reached. When V_o is increased
some more, I increases and a voltage drop appears. Along ab one
says that the system is in the resistive superconducting state
(R.S. state).

The interpretation of the behavior of the bridge along ab is
that vortices are created along the line NP. The structure of
these vortices consists of a nucleus in which is trapped a
quantum of magnetic flux, surrounded by concentric super
currents. This magnetic field is perpendicular to the film.
The Lorentz force between the magnetic fluxes and the current
induces the vortices to move sideways along the line NP.
(Figure 8)

It can be shown that there exists a simple relation between
the rate dn/dt at which the vortices cross the junction, and the
potential drop (6):

(11) $\dfrac{dn}{dt} = \dfrac{2ev}{h}$

Other measurements have been performed that show that the potential drop V is indeed due to a motion of vortices along NP.

In the language of the preceding chapters the PS state
corresponds to the steady state and the RS state to the dissipa-

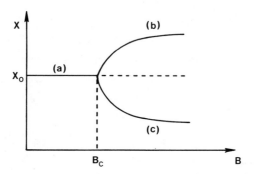

Figure 5.   Bifurcation diagram for the trimolecular model

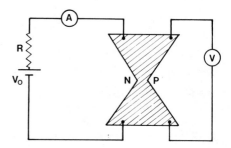

Figure 6.   Thin film superconducting
bridge

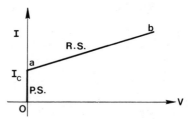

Figure 7.   Current–voltage characteristic for an Anderson–Dayem bridge

tive structures state.

(i) The PS state is analogous to a steady state because it corresponds to a uniform current flow through the bridge. Still it is different, because a steady state flow usually takes place away from equilibrium. In the superconducting case the super-current I does not dissipate any energy and could thus take place at equilibrium (the same supercurrent could flow indefinitely through the bridge if it had the shape shown in Figure 9.

(ii) The RS state corresponds to the dissipative structure state. Indeed new structures, a moving array of vortices, appear because the system is away from equilibrium. There is still an important difference from the usual dissipative structures, in that no energy is dissipated by the supercurrent of the vortices. No energy dissipation is necessary to keep the vortices alive. This is not the case for usual dissipative structures, such as the Bénard cells for instance, where the liquid flow dissipates energy.

Actually vortices similar to the one appearing in the bridge can be obtained at equilibrium by putting the film in a perpen-dicular magnetic field.

The energy dissipated in the RS state is used to move the array of vortices sideways.

Oxide Barrier Josephson Junctions (7,8)

A similar behavior is observed in the oxide barrier Josephson junctions. We can think of these junctions as formed by two superconductors separated by a thin insulator (Figure 10). Let us call

$$(12) \quad \psi_1 = \sqrt{\rho_1}\, e^{i\theta_1} \quad \text{and} \quad \psi_2 = \sqrt{\rho_2}\, e^{i\theta_2}$$

the electron wave function in superconductors 1 and 2, and $\phi = \theta_1 - \theta_2$. Josephson has shown that the macroscopic behavior of the barrier is described by the phenomenological equations:

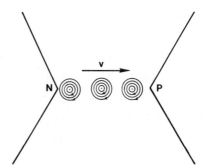

Figure 8. Array of vortices appearing
in an Anderson–Dayem bridge

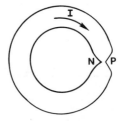

Figure 9. The current flowing without
energy dissipation in a thin film ring
illustrates the fact that in superconduc-
tivity one can have flow in an equilibrium state.

$$(13) \quad \frac{\partial \phi}{\partial y} = (\frac{2ed}{hc}) H_z \qquad \frac{\partial \phi}{\partial t} = (\frac{2e}{h}) V$$

$$J_x = j_1 \sin \phi + \sigma V$$

where d is the effective thickness of the barrier and H_z is the magnetic field inside the barrier along the z direction. Combining these equations with the Maxwell equations, one gets

$$(14) \quad [\frac{\partial^2}{\partial y^2} - \frac{1}{v^2} \frac{\partial^2}{\partial t^2} - \frac{\beta}{v^2} \frac{\partial}{\partial t}] \phi = \frac{1}{\lambda_J^2} \sin\phi$$

where $\quad \beta = \frac{4\pi d v^2 \sigma}{c^2} \quad$ and $\quad \lambda_J^2 = \frac{hc^2}{8\pi ed J_2}$

It is not difficult to show that equation (14) admits two different types of solutions, a time independent solution and a time dependent solution (the reader who wants to see this in detail should read the corresponding chapter of Ref. 8).

Both solutions correspond to arrays of pseudo vortices along the oxide barrier (see Figure 11).

In the time independent case the array is stationary and does not dissipate any energy; in the time dependent case it moves along the barrier with an arbitrary velocity $V^1 < V$. The moving array of vortices appears, as in the Anderson-Dayem bridges, when the current through the junction exceeds a certain critical current I_c. In that case the system is in a dissipative structure state.

We can make the same conclusions as in the case of Anderson-Dayem bridges. Away from equilibrium an array of vortices appears which moves along the barrier. No energy dissipation is needed to keep the vortices alive, the proof being that similar arrays (the stationary arrays) can be produced at equilibrium. The energy that is dissipated is used to move the vortices along the barrier.

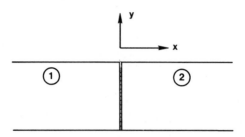

Figure 10.   *Oxide barrier Josephson junction*

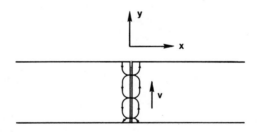

Figure 11.   *Array of pseudovortices along the barrier of a Josephson junction*

Conclusions

As we have seen, the Josephson effects appear to occupy an intermediate position between phase transitions and bifurcations to dissipative structure states. One of the reasons for this special behavior of superconductors is that they can carry electric currents without dissipating any energy. There is another reason for the Josephson effects to occupy this intermediate position. In the case of ordinary dissipative structures, the structures are macroscopic and can be described by classical physics.

In the case of superconductors, although the structures are macroscopic they have to be described by quantum mechanics. It is a unique property of superconductors to show quantum structures on a macroscopic scale. (For instance the magnetic flux trapped in the vortices mentioned in this section is given by the expression, $\phi_o = \dfrac{h}{2e}$. The dimension of the core of such a vortex is several hundred Angstroms.)

Literature Cited

1.   Glansdorff, P. and Prigogine, I.   "Thermodynamic Theory of Structure, Stability and Fluctuations"; Wiley-Interscience: New York, 1974).

2.   Nicolis, G. and Prigogine, I.   "Self-organization in Non Equilibrium Systems"; Wiley-Interscience: New York, 1977).

3.   Prigogine, I.   "Introduction to Thermodynamics of Irreversible Processes"; Wiley-Interscience: New York, 1967).

4.   Haken, H.   "Synergetics, An Introduction"; Springer-Verlag: New York, 1977).

5.   Dayem, A. H.   Phys. Rev., 1967, 155, 2, 419.

6.   Anderson, P. W.   Progress in Low Temperature Physics, 1967, Volume V, Chapter I.

7.   Stephen, J.   Phys. Rev., 1967, 163, 2, 376.

8.   Solymar, L.   "Superconductive Tunnelling and Applications"; Chapman and Hall, Ltd.: New York, 1972.

RECEIVED September 25, 1979.

Clarification and Obfuscation in Approaching the Laws of Thermodynamics

FRANK C. ANDREWS

Merrill College, University of California, Santa Cruz, CA 95064

A statement of a "law" in science is a generalization from experience -- one whose logical consequences are so sweeping as to correlate, make predictable and "understandable" a large enough range of phenomena to justify a designation as presumptuous as the word "law." Natural phenomena can be analyzed in various ways; the statements we now call scientific laws are artifacts of the history of scientific thought. There is no "right" statement of a scientific law; different contenders can be judged at a given moment on the basis of clarity and utility in learning and extending understanding of the phenomena in question.

The laws of thermodynamics have attracted more attention to their formulation than many other scientific laws, perhaps because of the self-contained nature of thermodynamic phenomena and the "beauty" of deriving so incredibly many (valid) results from such seemingly paltry starting material. Also, those who study thermodynamics commonly experience it as one of the hardest subjects they take; it is abstract, its problems challenge their algebraic thinking to the fullest, problems rarely repeat types that were seen before, but they are ever-new and unexpected, thermodynamics is constantly being applied to new phenomena and in different ways. Thus, students experience delayed understanding of thermodynamics, often not mastering it until they have themselves taught or applied it for many years. There is then a temptation to seize on the way it is finally understood as the "right" way, the "best" way, the way that "if only I had heard that years ago, I would have been spared all this puzzled misunderstanding." We neglect the importance of all those years of work in our own understanding. So we all write our books and offer our formulations as if the way our approaches differed from each other were really very important. And still our students have trouble understanding us whatever book we use, however we state the thermodynamic laws. So we go on hunting for the "right" statements, the ones that will make thermodynamics transparent quickly to everyone who wants to learn. It will be a long hunt.

0–8412–0541–8/80/47–122–245$05.00/0
© 1980 American Chemical Society

We shouldn't get carried away by our deductive disciplines. From our own sophisticated viewpoints, within the context of understanding we already have when we approach, say, thermodynamics -- the deductions appear clean and logical to us. That's because we take for granted so much that has to be invoked tacitly to develop the results. Our students, who cannot put themselves in our sophisticated shoes, don't know what they're supposed to take for granted. Derivations are plausibility arguments. For a given result to be "derived," a minimum amount of physical content must be built in at the start or along the way. Why don't we simply concentrate on what that minimum amount of physics is -- perhaps show the different equivalent ways to package that minimum -- and avoid fixation on the "best" way to build it in?

Let us spend a few minutes considering what minimum physical content is necessarily involved and must be present in any formulation of the laws of thermodynamics. In the course of this, a number of conventional statements of the laws will crop up. Later, we'll list some criteria by which to evaluate statements of the laws and apply them to various statements.

Thermodynamics works only for systems described by macroscopic properties. Such properties, even the intensive ones, must involve huge numbers of the particles that make up the system. They must also involve measurement times long compared to the durations of significant fluctuations among the particles involved in the measurement.

The state of the system must be capable of being specified by a limited number of independent thermodynamic properties. In a deterministic macroscopic discipline (like thermodynamics), the state is specified when any system constructed according to the specification will evolve subsequently in the same way. This means there must be no hidden variables; all such must be identified and brought to light. A pile of oily rags in a corner of the room gets hotter and hotter until it spontaneously bursts into flame. So we identify the progress variable for the chemical reaction of oil with oxygen as a property of the system that we otherwise would have overlooked. A lump of uranium stays warmer than the room over a period of time. So we identify the fraction of it that has decomposed radioactively as a property, previously overlooked, of the system. A Mexican jumping bean suddenly leaps off the table. So we presume it contains either an internal spring or some other hidden property which we must know in order to specify its state. Specifying the state of a living organism in the sense given here is impossible and of course thermodynamics is not what determines the behavior of living creatures.

The ways in which different ones of us lay the groundwork for the first law can be best viewed in light of our objective, which is always the equation

$$\Delta E = Q \pm W. \tag{1}$$

Here, E is the system's energy, Q the heat absorbed by the system,
and W the work done on (+) [or by (-)] the system by [on] the
surroundings during the process of interest. We want each of
these three quantities to be unambiguous, intuitive, and seen as
measureable, at least in principle. Since we're never sure
where our students are in their understanding of these three
quantities, we never quite know where to start.

In basic physics, the existence of energy as a function of
state has its origin in the reproducibility of nature, the fact
that the laws of nature are independent of the origin of time.
In elementary mechanics, work establishes the function of state,
energy, whose existence proves so useful in practical problems.
In elementary mechanics there is no heat, friction, or hidden
variable like temperature, chemical reaction, or radioactive
decay. In this idealized world, work is completely convertible
from one form to another (e.g., electrical to gravitational),
and we can express each form mathematically in terms of related
properties of the system. Thus, in mechanics, if one takes a
system from state A to state B with the sole effect in the
environment being any unidimensional measure of the total work,
that measure will be independent of path between A and B. An
example of such a measure is the change of height of a standard
weight in a gravitational field. In mechanics this change will
depend solely on states A and B and not on which process took
the system from A to B. Thus, the change in height of the weight
measures the change in a function of state, the energy.

Now suppose we take a real thermodynamic system and introduce the concept of adiabatic and diathermal walls in order to
distinguish between work and heat. For the moment we retain the
system within adiabatic walls so only mechanical interactions
can occur between it and its surroundings. However, we allow
non-mechanical thermodynamic interactions inside the system
itself. The same result is observed as in mechanics: for a
given change in state of the system, the energy change in the
surroundings is unique. Specifying states of the system requires
thermodynamic and not just mechanical properties. These are such
things as temperature, extent of a chemical reaction, and extent
of a radioactive decay. But any unidimensional measure of energy
change in the surroundings will still read the same for the two
states of the system, independent of the path taken between them.
This generalization from experience contains the gist of the
first law and permits use of the energy as a function of state
in thermodynamic systems.

With this approach, the heat Q enters almost as a "fudge
factor" in a non-adiabatic process, $Q \equiv \Delta E \mp W$. But the interesting
point is that heat defined in this way is precisely the heat of
our everyday experience. This is the remarkable result of

Rumford's cannon-boring and Joule's paddle-wheel experiments that
led up to the first law. Indeed, heat and work are parallel
concepts in thermodynamics. Students, depending on their back-
grounds, may feel more at home with one or the other. Measures
of work are more direct, usually, than measures of heat. However,
we could have chosen for our unidimensional energy measure in the
surroundings the (empirical) temperature change in a standard
heat reservoir just as well as the change in height of a standard
weight. The close relationship between the existence of energy
as a function of state and the reproducibility of nature is
suggested by either formulation of the first law. If we sum all
the change in the surroundings into a single property whose
change is path independent, that property can be mechanical and
the walls adiabatic, in which case the first law is $\Delta E = W$ for
adiabatic processes. That property could be thermal and the
walls diathermal, in which case the first law is $\Delta E = Q$ for
systems with fixed constraints. Either way, the result is the
same, $\Delta E = Q \pm W$. In fact, there was much merit in stating the
first law in the form of the impossibility of a perpetual motion
machine of the first kind, one which could undergo a cycle and
leave no other effect than a different value for whichever unidimensional measure of energy in the surroundings one chose,
work or heat.

However we choose to state the second law of thermodynamics,
the gist of it is that the systems left to themselves run down
and reach equilibrium. The macroscopic properties of any isolated system eventually assume constant values. This is a
statement about the real world which is valid only for macroscopic thermodynamic properties. Ever-present fluctuations in
even equilibrium intensive properties represent violations of
the law as stated. Predictions made from the law will be of the
average equilibrium properties; useful study of fluctuations
demands a microscopic theory. So far as we know, this statement
of the second law, that systems run down and reach equilibrium,
is true of all processes within our solar system and perhaps
our galaxy, maybe even of our universe. The statement is
intuitive; all of us have seen many examples of systems reaching
equilibrium, and people can check the statement by such means as
putting a mouse in a bell-jar or an ice cube in some warm water.
This statement has the advantage of discouraging the search
for perpetual motion machines (of the second kind) that attracts
so many people whose understanding of the second law is based on
other statements that are less intuitive. The statement is
relaxed and friendly, and conveys even to the non-scientist the
important qualitative truth generalized in this fascinating law.
Now, suppose we want to establish the whole quantitative
framework of thermodynamics based on the second law. We must
first attend to temperature. We want to conclude that temperature can meaningfully be taken as a property of a system and

that numbers can consistently be assigned to temperature so that
hotter systems always have higher temperatures than colder, i.e.,
that spontaneous heat flow is driven by temperature difference.

We want to assign a common temperature to all systems that
would prove to be in thermal equilibrium with each other. Thus
we must be sure that this process would always prove consistent.
We need to know that <u>things in thermal equilibrium with the same
thing will always be found in thermal equilibrium with each
other</u>. This can be evoked as the zeroth law of thermodynamics.
Or it can be derived as a consequence of the statement of the
second law chosen above, if one is willing to grant the existence
of heat engines. Heat engines are devices having heat-only
interactions with two objects, absorbing heat from one, discharg-
ing heat to the other, and spontaneously doing work (moving
around, e.g.) in the process. A thermocouple is a good example.
Heat engines are a class of diathermal walls that can wiggle as
heat flows through them. Given some diathermal walls that
wiggle, it is clear that every conceivable kind of diathermal
wall, especially including even the most sensitive wiggling ones,
must agree on what are states of thermal equilibrium. Otherwise
a system could be constructed that wiggled forever and thus
violated the second law. The zeroth law is thus a special case
of the second law which stresses thermal equilibrium and prevents heat engines from being used to create perpetual motion.

This is not enough, by itself, to assign numbers meaningfully to temperature, though most books would have you believe it
was. We must go further and establish that if A is hotter than
B in one experiment, it will always be hotter than B in all
other experiments. By hotter, we mean if we put A and B into
thermal contact, heat flows spontaneously from A to B. If I
have done one experiment and found A to be hotter than B, I want
to show that <u>it is impossible to have a natural process which
produces no other effect than the absorption of heat from the
colder body</u> (B) <u>and the discharge of that heat into the warmer
one</u> (A). This is Clausius' statement of the second law, and it
is needed for establishing a temperature scale. It follows
directly from my statement of the second law as the inevitability
of equilibrium, granted the existence of heat engines. If it
weren't true, heat could flow through a wiggling diathermal wall
from A to B, and the energy could be returned to A from B via
the Clausius' law violator. The result would wiggle indefinitely.

So, we can assign temperatures consistently with our experience of hotness. Now it's time to move on and prove both the
existence of entropy as a state function and the Clausius
inequality. This requires employing the concept of the reversible
process. A reversible process is one each step of which may be
exactly reversed by an infinitesimal change in the external
conditions prevailing at the time of that step. The two requirements needed for a real process to approximate reversibility are
(1) the process proceeds slowly compared to all internal

relaxation times of importance, and (2) there must be no finite
temperature differences, friction, overvoltage, hysteresis, etc.
At any moment in a reversible process the intensive properties
involved in the surroundings and the system will be identical.

It is convenient to note Kelvin's statement of the second
law: It is impossible to have a natural process which produces
no other effect than the absorption of heat from a single reservoir and the performance of an equivalent amount of work. This
is obviously a consequence of the inevitable approach to
equilibrium, since if Kelvin's statement were violated, the work
could be done back on the reservoir while something spontaneously
moved, and we would have a device that could wiggle forever.
Kelvin's statement stresses how hard it is to turn random thermal
energy of a heat flow into organized mechanical energy of work.
The reverse, turning work into heat, is of course no problem at
all.

Now suppose we represent the state of our system by the
values of the constraints (volume V and others x) and internal
energy U. Imagine the system represented by a point in a 3-
dimensional space with V and x values as the horizontal axes and
U as the vertical. It is not trivial that we can so represent
the system; the inevitable approach to equilibrium of any system
so constrained and started with so much energy assures us that
we can. Now suppose we start at some arbitrary point and change
the system reversibly and adiabatically by varying the constraints, doing work, but not allowing any heat to flow. We can
immediately conclude that such changes can never take us to a
state with the same constraints but lower U than the initial
state. Work would be done by the system that ends with its
initial constraints. If they did, we could let the system at
those fixed constraints then absorb enough heat from a single
hot reservoir to restore it to its original U. It then would
have undergone a cycle which violated Kelvin's statement. If
reversible adiabatic paths cannot take us to the same constraints
and lower U, they can't take us to the same constraints and
higher U either, since the reverse of a reversible path is possible by definition. Thus, reversible adiabats must form a
non-folding surface in our U-V-x space, and we call these
surfaces of constant entropy. This is clearly what Carathéodory
had in mind when he gave us what must surely be one of the most
obfuscating statements of the second law: in the neighborhood
of any prescribed initial state, there are states which cannot
be reached by an adiabatic process. These unreachable states are
those with the same V and x but lower U. Carathéodory's statement follows from the inevitability of equilibrium. Its usefulness is in clarifying why entropy is a state function.

Now if we change from one reversible adiabat to another
nearby by reversibly absorbing heat, it is straightforward to
show that

$$\mathrm{d}Q_{rev} = TdS \qquad \text{where} \qquad T \equiv \left(\frac{\partial U}{\partial S}\right)_{V,x}. \qquad (2)$$

This T, the thermodynamic or absolute temperature, is here a function of S, V and x. But it's easy to show that if T were a function of temperature and entropy, or if it were a function of temperature and anything else, we could violate Kelvin's statement. So T depends only on the empirical temperature, and this dependence must be the same for all systems in order for the entropy of a composite to equal the sum of the entropies of the subsystem. In order for Clausius' statement to hold in the case of irreversible processes, the equal sign of $\mathrm{d}Q_{rev} = TdS$ becomes <, and we have Clausius' inequality: $\mathrm{d}Q \leq TdS$, where T is the thermodynamic temperature, a particular universal function of only the empirical temperature.

One point that doesn't seem to be stressed enough to students is why the entropy is so extremely useful. We have emphasized that there is a unique equilibrium state for the system with specified U, V, and x. That being the case, the intensive equilibrium properties like temperature and pressure must be functions of U, V, and x. But what functions? Must we simply tabulate enormous quantities of data, or devise empirical equations to interrelate all the many properties? Here is where the entropy offers the dramatic simplification that gives thermodynamics much of its power. The entropy is a <u>thermodynamic potential</u> from which the relations among the various system properties can be found by differentiation. If we know the single function S = S(U,V,x), then $1/T = (\partial S/\partial U)_{V,x}$ and $p/T = (\partial S/\partial V)_{U,x}$. We can take Legendre transforms of the entropy or energy to create other thermodynamic potentials whose natural independent variables are more convenient than those natural to the energy or entropy. Thus arise the familiar enthalpy and free energies. Data from whatever sources, most commonly p-V-T data and values of heat capacities and latent heats, can be used to maximum advantage in predicting all kinds of other data. Students rarely appreciate the powerful simplification this introduces until they have the experience in statistical mechanics of obtaining the partition function and from it the Helmholtz free energy, and from that all the rest of the thermodynamic relationships.

Now, if we contrast thermodynamics with statistical mechanics, we find that they confront completely different problems. The problem of statistical mechanics is to compute the probabilities of the various possible results of a variety of measurements one might choose to make on a system, (a) knowing it consists of particles with known mechanical properties, and (b) knowing certain macroscopic information about the system. The problem is approached by building the appropriate information into an ensemble which represents the system and generates all

the probabilities of interest. The basic assumption of equilibrium statistical mechanics is that the only dynamical feature on
which the probability of a state may depend is the energy of the
state. In the case of an isolated equilibrium system, this
assumption leads to a microcanonical ensemble. Another way to
view this ensemble is that it contains a minimum amount of information beyond the measured energy of the isolated system, where
information is measured as in information theory. This assignment of equal probabilities to all states of the same energy is
plausible and can be made more plausible by arguments from both
statistical mechanics and information theory. But in the long
run the assignment is arbitrary, and comparison of results with
experiment is the final check of its usefulness.

In statistical mechanics the existence of energy as a
function of state follows from the existence of energy in mechanics. Each member of the ensemble is, after all, being treated
as a completely mechanical system. Indeed, one problem of
statistical mechanics is finding ways to calculate the non-
mechanical thermodynamic quantities like heat. This can be begun
directly by differentiating the ensemble average (predicted
value) of the energy, $\overline{E}$, for some arbitrary reversible process:

$$d\overline{E} = d(\Sigma_i P_i E_i) = \Sigma_i P_i dE_i + \Sigma_i E_i dP_i = đW + đQ. \qquad (3)$$

Here, P_i is the probability that the system be in the ith N-
particle quantum state, E_i is the energy of quantum state i, and
the sum is over all quantum states for the N-particle system.
The resemblance of Eq. 3 to the first law permits some valuable
identifications. The sum over $P_i dE_i$ is identified as the average
value of the work done on the system; it is the average over an
unchanged ensemble of the changes in energy eigenvalues of all
the states caused by the varying constraints, such as volume.
The sum over $E_i dP_i$ is the average value of the heat; it is the
energy needed to redistribute the probabilities among the various
quantum states caused by the process being considered.

From statistical mechanics the second law as a general
statement of the inevitable approach to equilibrium in an isolated system appears next to impossible to obtain. There are so
many different kinds of systems one might imagine, and each one
needs to be treated differently by an extremely complicated non-
equilibrium theory. The final equilibrium relations however
involving the entropy are straightforward to obtain. This is not
done from the microcanonical ensemble, which is virtually impossible to work with. Instead, the system is placed in thermal
equilibrium with a heat bath at temperature T and represented by
a canonical ensemble. The presence of the heat bath introduces
the property of temperature, which is tricky in a microscopic
discipline, and relaxes the restriction that all quantum states
the system could be in must have the same energy. Fluctuations
in energy become possible when a heat bath is connected to the

system. This permits factoring the system probability into terms
representing different parts of the system, and the result is
the canonical ensemble. For this ensemble it is easy to show
that the reversible heat has an integrating factor which can be
called the temperature, and the entropy can be identified as (-k)
times the ensemble average of the natural log of the probability
of a state. The k here can be identified as Boltzmann's constant,
the universal gas constant on a per molecule basis, only after
statistical mechanics is applied to ideal gases and the result
compared to experiment. Boltzmann's relation, S = k ln W,
follows for a system in which there are W equally probable states
as a limiting case of the canonical ensemble.

Thus, statistical mechanics' development of the thermodynamic laws is interesting and straightforward enough, and it
sheds helpful light on thermodynamics. So is the use of information theory to develop and discuss statistical mechanics. They
supplement the understanding generated by the macroscopic statements, but in no sense do they replace that understanding.
Based on my own teaching and studying experience, I favor giving
the qualitative molecular picture along with macroscopic thermodynamics and saving the quantitative molecular picture for a
separate treatment of statistical mechanics. Anything that can
help in making thermodynamics more intuitive is a blessing, and
pictures of molecules bouncing around and interacting with each
other is of great intuitive help. Some people prefer a more
thorough mixing of the two theories, but I find the quantitative
study of either to be hard enough without the other.

One aspect of thermodynamics in which statistical mechanics
is especially helpful is the third law. The third law of thermodynamics is a funny thing to have acquired the august name of a
"law." This probably happened at a time when people were enrap-
tured with the beauty of deriving all of thermodynamics from a
handful of postulates called laws. The first two laws were given
early and deserved stress, and then it was realized that further
input was needed to get everything, to make the logic tighter.
So the third law and zeroth law were named. Now it seems clear
that a lot of other input is needed to get the entire structure
of thermodynamics and apply it in the real world. Let's hope
people have given up numbering laws.

There are two common statements of the third law, which are
sometimes claimed to be logically identical, but which probably
are not. The first is that absolute zero can never be reached,
even in theory. This certainly seems plausible as a practical
matter because the system is bound to absorb energy from the
radiation field emitted by whatever surrounds it. On the other
hand, this law seems of no importance, aside from destroying our
dreams of making a heat engine 100% efficient by using a cold
reservoir at absolute zero. This reservoir would never have to
warm up because a 100% efficient heat engine dumps no heat into
its cold reservoir. Theoretical unattainability of zero Kelvin

in unimportant as a practical matter because temperatures of 10^{-4} and 10^{-5}K are routinely obtained, and no dramatic change in material properties is likely to occur between zero and 10^{-5}K.

The second statement of the third law (which bears Planck's name) is that as the temperature goes to zero, ΔS goes to zero for any process for which a reversible path could be imagined, provided the reactants and products are perfect crystals. Here, perfect crystals are defined as those which are non-degenerate, that is, they have only a single quantum state in which they can exist at absolute zero. This statement follows rigorously from Boltzmann's equation for entropy,

$$S = k \ln W = k \ln 1 = 0. \tag{4}$$

Unlike enthalpies for which values at absolute zero are important, hard to measure, and must be tabulated carefully, entropies of most pure compounds can be taken to be zero at absolute zero. Tabulations of entropy are thus simpler than tables of enthalpies, and finding entropy changes for chemical reactions from tables is often more precise than finding the comparable enthalpy change because there is no comparable law for enthalpies.

Three points are worth making about this third law. First, if it weren't true, it would not be a big deal. We would just have to tabulate entropies at absolute zero as we already do for enthalpies and forego the expression "absolute entropy." Second, some compounds have "residual entropies" at absolute zero as it is, and we can cope with and understand them easily. Third, there are no perfect crystals, there never will be, and there don't need to be in order for the entropy at absolute zero to measure zero. What is the maximum experimental precision of an entropy measurement? Perhaps 10^{-5} eu? How big can the degeneracy of a crystal be before its entropy becomes 10^{-5} eu?

$$10^{-5} \text{ eu} = k \ln W = (3.3 \times 10^{-24} \text{ eu}) \ln W$$

$$\ln W = 3 \times 10^{18}.$$

Thus, a crystal is effectively perfect so long as its degeneracy is less than about $10^{10^{18}}$! So, our imperfect real crystals can have innumerable lattice imperfections, dislocations, fractures, and chips and have zero residual entropy, so long as "innumerable" in this case does not mean more than $10^{10^{18}}$. Since $10^{10^{18}}$ is such a large number, we can see why it requires some element of degeneracy associated with each molecule or few molecules in order for the residual entropy of a crystal to be measureably large.

Let me end by listing some criteria by which to evaluate statements of thermodynamic laws. I'll include comments about how some of the commoner statements fare when judged by each criterion. This is an exercise we might encourage our students

to do for themselves, thus showing that we are not hung up on
which statement they learn (for parroting back purposes), so long
as they are aware of its implications and make it work well for
themselves.

 1. How true is the statement?   The first law stated in the
form "the energy of the universe is constant," for example, is
currently believed by cosmologists. But only a few years ago
when continuous creation rivalled the big bang theory, many
believed it was false. Surely something as important as the
first law of thermodynamics shouldn't be stated in a form whose
validity shifts with the latest developments in the turbulent
discipline of cosmology. The same can be said for the analogous
statement of the second law, "the entropy of the universe evolves
to a maximum." It is quite possible that we live in an oscil-
lating universe which will return to its original superdense
state at some low-entropy time in the future.

 2. How subject is the statement to experimental checking?
Any statement invoking the universe surely fails on this score,
too, since how can one experiment with the universe? Why do we
drag the universe into statements about experiments done only on
our earth? Statements out of statistical mechanics and information theory often rest on postulates and mechanical details many
steps removed from experimental verification. All statements
about something being impossible such as Kelvin's and Clausius'
are hard to imagine checking. Stating a law in the form of an
impossibility seems to emphasize the baldness of an inductive
generalization.

 3. How intuitive is the statement?   How accustomed are
students to the properties and phenomena involved in the statement? How directly does the statement suggest the evidence for
and consequences of the law? Here is where most statements of
the second law fall far short of what is desired. Consider
Carathéodory's, "In the neighborhood of any prescribed initial
state, there are states which cannot be reached by an adiabatic
process." Is this really a good way to express what is one of
the most profound and fruitful generalizations in all of science?
Again, the popular impossibility statements of Kelvin and of
Clausius seem to be relatively minor truths, fine perhaps as
corollaries, but intuitively not very obvious. The postulates
of statistical mechanics and information theory are separated
from our every-day intuition by enormous gulfs.

 4. How well does the statement avoid introducing technical
concepts that themselves require extensive discussion?   For
example, does a statement of the first law mention energy? Does
a statement of the second law mention entropy? With no obvious
way to measure such concepts, we must resign ourselves to lengthy
explanations. Concealed somewhere in those explanations the
actual content of the law at hand is most likely to be introduced.
Indeed, property, state, work, heat, energy, equilibrium, temperature, reversibility, entropy, probability, and information are

such precise, technical, loaded concepts whose proper treatment requires so much sophistication, that their use in stating the thermodynamic laws is unfortunate. True, much of thermodynamics consists of elucidating these concepts. That's all the more reason for avoiding them in the most basic statements.

5. How similarly does the statement treat similar concepts? This is as opposed to seizing on only one concept to make much of and thus distorting the fact that the concepts are similar. For example, stating the first law as "the work done by an adiabatic system is a function of the initial and final states only and not of the path," singles out work for prominence. Later we slip the heat in as a correction term to preserve energy conservation in non-adiabatic systems. This approach conceals the important parallel between heat and work, it distorts perspective for the sake of proving a point.

6. How relaxed and pleasant a subject does the statement suggest thermodynamics to be? I'll let each of you think back to some of the statements of the thermodynamic laws you've read (or written?) to come up with examples that shout to the reader that this will be the most tortured exercise in nit-picking he or she has run across in all of science.

I suggest that we put more effort into really understanding thermodynamics and less into starting a pedagogical revolution by restating a law. The better a person understands a subject, the easier is that person to learn from, the quicker and more accurate will that person's thinking be. Thermodynamics is hard for most students, important in the real world, and inadequately understood by the average scientist. The payoff in both fun and utility awaits us in understanding the whole subject better than we do now, through thousands of little points cleared up, through lots of hard work.

General References

Andrews, F. C., "Thermodynamics -- Principles and Applications," Wiley-Interscience, New York, 1971.

Andrews, F. C., "Equilibrium Statistical Mechanics," Wiley-Interscience, New York, 2nd ed., 1975.

RECEIVED October 29, 1979.

The Laws of Thermodynamics:
A Necessary Complement to Quantum Physics

ELIAS P. GYFTOPOULOS

Massachusetts Institute of Technology, 77 Massachusetts Ave.,
Cambridge, MA 02139

GEORGE N. HATSOPOULOS

Thermo Electron Corporation, 101 First Ave., Waltham, MA 02154

The purpose of this paper is to discuss the hypothesis that the second law of thermodynamics and its corollaries are manifestations of microscopic quantum effects of the same nature but more general than those described by the Heisenberg uncertainty principle. This hypothesis is new to physics.

Energy is a property of matter that has unified our understanding of physical phenomena. It appears prominently in both mechanical and thermodynamic theories. It is defined by the first law of thermodynamics, and its conservation is one of the keystones of analyses of both microscopic and macroscopic phenomena.

Adiabatic availability--the amount of work that can be extracted from a system adiabatically--is also a property of matter but has not been accepted with the same unanimity as energy. Experience, especially with large bodies, indicates that not all the energy of a system is available in the form of work. For a given system, say atoms in a box, mechanics implies that all the energy (above the ground state) can be extracted adiabatically in the form of work, regardless of whether the system is large or small. In contrast, the second law of thermodynamics implies that not all the energy can be extracted in the form of work, namely that the adiabatic availability is smaller than the energy and that it can be even zero.

Though perhaps expressed in different terms, the contrast just cited has been the subject of intensive inquiry and controversy ever since the enunciation of the first and second laws of thermodynamics in the 1850's. Invariably, a reconciliation is proposed based on regarding thermodynamics as a statistical macroscopic or phenomenological theory.

We believe that such reconciliation is not adequate. Moreover, we believe that the thermodynamic behavior of matter is due to quantum uncertainties of the same nature but broader than those associated with wave functions and invoked in the uncertainty principle.

In what follows, we discuss the current interpretation of the relation between thermodynamics and microscopic structure of

0–8412–0541–8/80/47–122–257$05.00/0
© 1980 American Chemical Society

matter, the accepted representation of quantum uncertainties or
dispersions, the proposed generalization of quantum dispersions,
and an inconsistency that led us to introduce this generalization.
Finally, we outline a quantum theory that encompasses both mechanics and thermodynamics without inconsistencies.

Statistical Thermodynamics

In the interest of brevity, we will present our discussions
in the language of quantum mechanics. Though many phenomena encountered in mechanics and thermodynamics can be expressed in
terms of classical mechanics, the central point we wish to make
cannot.

The dominant view currently held about the physical significance of thermodynamics is based on the interpretation of a
"thermodynamic state" as a composite that best describes the
knowledge of an observer possessing only partial information about
the "actual state" of the system. The "actual state" at any instant of time is defined as a wave function (a pure state or a
projection operator) of quantum mechanics. The theories that have
recently evolved pursuant to this view have been called informa-
tional, though the same concept is the foundation of all statistical thermodynamics.

In such theories the partial or uncertain knowledge of the
observer is represented by probabilities, each of which is assigned to one of the possible "actual states." The assignment is
achieved by using a postulate or a hypothesis in addition to the
laws of mechanics. Recently, the most prevalent postulate is
Shannon's criterion. (1, 2, 3, 4) It defines the measure of uncertainty of the observer about the state of the system in terms
of the probabilities assigned to the possible "actual states."

The value of the measure just cited ranges from zero to some
conditional maximum depending on whether the observer is completely informed or possesses incomplete information about the
"actual state" of the system, respectively. Whatever its value,
this measure of uncertainty of the knowledge of the observer is
identified with the entropy of thermodynamics. Thus, entropy is
perceived as an entity characteristic of the knowledge of an observer rather than as an inherent property of matter. In other
words, statistical entropy does not have the same physical under-
pinning as do energy, mass, and momentum.

The interpretation of entropy as a measure of knowledge of an
observer cannot be avoided in any of the statistical theories of
thermodynamics advanced to date because in all these theories objective reality is represented by pure states only, and such
states have no entropy.

Following von Neumann, the "thermodynamic state" is expressed
in terms of a statistical matrix or statistical operator that includes combinations of two kinds of probabilities, those that describe the uncertainties characterizing the knowledge of the

observer and that are obtained from the hypothesis added to the
laws of mechanics, and those that describe the uncertainties inherent in the nature of matter and that are obtained from the alternative wave functions of the possible "actual states" under
consideration. It is a mixed state that consists of a statistical
mixture of pure states.

Limited Quantum Uncertainties

In classical mechanics it is assumed that at each instant of
time a particle is at a definite position x. Review of experiments, however, reveals that each of many measurements of position
of identical particles in identical conditions does not yield the
same result. In addition, and more importantly, the result of
each measurement is unpredictable. Similar remarks can be made
about measurement results of properties, such as energy and momentum, of any system. Close scrutiny of the experimental evidence
has ruled out the possibility that the unpredictability of microscopic measurement results are due to either inaccuracies in the
prescription of initial conditions or errors in measurement. As
a result, it has been concluded that this unpredictability reflects objective characteristics inherent to the nature of matter,
and that it can be described only by quantum theory. In this
theory, measurement results are predicted probabilistically,
namely, with ranges of values and a probability distribution over
each range. In contrast to statistics, each set of probabilities
of quantum mechanics is associated with a state of matter, including a state of a single particle, and not with a model that
describes ignorance or faulty experimentation.

Measurement results and their probabilities can be used to
compute the standard deviation or dispersion of the results. If
the dispersion in measurements of position x is denoted by Δx, and
of momentum p_x along the x-axis by Δp_x, it is found that the product of Δx and Δp_x is greater than a lower limit equal to Planck's
constant. This inequality is the well known Heisenberg uncertainty principle. We will see later that other measures of uncertainty or dispersion are possible.

It is customary to represent the probabilities of a quantum-
mechanical or "actual" state by a wave function or, equivalently,
by a density matrix that is pure or idempotent, i.e., by a matrix
that, in diagonal form, has one element equal to unity and all
others equal to zero.

Broad Quantum Uncertainties – A Hypothesis

We will hypothesize that quantum conditions exist for which
the objective probabilities, inherent to the nature of matter,
must be represented by a density matrix that is not idempotent,
i.e., a matrix that, in diagonal form, has more than one element
different from zero. We will say then that we have objective

uncertainties or dispersions broader than those implied by idempotent density matrices, and we will identify such uncertainties
with the thermodynamic state. We will call such matrices mixed.

The hypothesis, and its identification with the thermodynamic state, raises several questions:

(1)   Are broader dispersions consistent with the laws of
      quantum mechanics?

(2)   Can the statistics represented by a non-idempotent or
      mixed matrix be reduced to a combination of statistics
      represented by idempotent or pure matrices?

(3)   Do broader dispersions introduce limitations other than
      those of conventional quantum theory?

(4)   Are broader dispersions consistent with the second law
      of thermodynamics?

We elaborate on the answers later in our presentation. Here
we wish to emphasize that, if our hypothesis is valid, it has the
following consequences:

(1)   The relations between expectation values of properties
      and the density matrix will be the same as in statisti-
      cal thermodynamics and, therefore, will conform to the
      laws of thermodynamics.

(2)   Because we consider one type of uncertainty only, namely
      the uncertainty inherent to the nature of matter, the
      laws of thermodynamics must complement the other laws of
      physics and must apply to all systems (large or small)
      and to all conditions (nonequilibrium, equilibrium, and
      stable equilibrium).

(3)   Availability and its related measure entropy become ob-
      jective properties of matter and cease to represent the
      extent of knowledge of observers. For example, a single
      particle has an entropy associated with its state in the
      same sense that it has energy and momentum associated
      with that state.

(4)   The breadth of dispersion is measured by entropy, which
      ranges in value from zero for a pure state (idempotent
      density matrix) of a given energy, to a maximum for the
      stable equilibrium (thermodynamic equilibrium) mixed
      state of the same energy.

An Inconsistency of Statistical Thermodynamics

Our motivation for advancing the hypothesis just cited can be
summarized as follows.

Statistical theories of thermodynamics yield many correct and
practical results. For example, they yield the canonical and
grand canonical distributions for petit and grand systems, respectively; these distributions, which were proposed by Gibbs, have
been shown by innumerable comparisons with experiments to describe
accurately the properties of quasistable states. Again, they pre-
dict the equality of temperatures of systems in mutual stable
equilibrium, the Maxwell relations, and the Gibbs equation.

Despite these successes, the premise that the thermodynamic state is a subjective characteristic of a partially informed observer rather than an objective characteristic of matter is questionable because it leads to an inconsistency.

If the thermodynamic state is subjective, then entropy and, therefore, availability are not objective properties of matter. As such, their values can be modified through improved knowledge acquired by means of some measurements. To account for this conclusion without contradicting the requirements imposed by the second law of thermodynamics, Szillard (5), and later on Brillouin (6) and others, advanced the view that acquisition of knowledge requires expenditure of work. Then they showed that increase in availability (or decrease in entropy) is more than offset by the work expended in obtaining better knowledge through measurement.

This view does explain why no work can be obtained out of a system in a stable equilibrium state. However, it does not explain why work can be obtained out of other systems not in stable equilibrium states even by observers totally ignorant about the "actual states" of the systems.

For example, take a collection of identical homogeneous tanks of water, all of which have the same energy, but some of which are isothermal in stable equilibrium states and some, by virtue of gradients in temperature, are not. An observer knowing nothing except the common nature of each tank and the common value of its energy could readily and reproducibly identify those tanks that are not in stable equilibrium states (as opposed to those that are) while expending on the average very much less work than the work that he can extract from them. In other words, it is an experimental fact that every observer can get a net amount of work out of a nonisothermal tank of water without knowing beforehand that it has temperature gradients.

Again, we know from experience that a system in a nonstable state usually, but not necessarily, proceeds towards a stable equilibrium state when isolated. For example, an internally stressed plastic solid may proceed, while isolated, towards an unstressed state at some rate, fast or slow. During such a process, some readily observable chracteristics of the system, such as its temperature, change with time. The changes occur in the system rather than in the mind of the observer, or as a result of interventions by the observer. Although the energy of the system has remained unchanged, the maximum value of the energy that can be transferred adiabatically to outside systems has decreased. In other words, the adiabatic availability of the system has changed and there is nothing any observer can do to restore it to its original value short of transferring availability from other systems.

These and other more sophisticated examples have a clear physical implication: departure from stable equilibrium is an inherent characteristic of a system that observers can ascertain reproducibly regardless of their personal knowledge about the system and, at least for large systems, with negligible work penalty. Otherwise we would not be able to exploit such energy resources as oil wells and uranium deposits.

In thermodynamics, for a given energy a measure of departure from stable equilibrium is the adiabatic availability--the larger the availability, the greater the departure from stable equilibrium. This availability in turn is related to the value of the entropy of the system since the smaller the entropy, the larger the availability. In other words, common sense dictates that entropy is an inherent property of matter in the same sense that energy is an inherent property of matter. Hence, the inconsistency of statistical thermodynamics.

The inconsistency is eliminated by the introduction of the hypothesis of existence of broad quantum dispersions discussed earlier. This hypothesis allows the unification of mechanics and thermodynamics. The unified theory was formally presented in a series of papers (7) in the "Foundations of Physics," and is briefly summarized below.

The Unified Theory

General Remarks. In this section, we will briefly outline an axiomatic theory that encompasses within a single quantum structure both mechanics and thermodynamics. A single quantum structure is necessary because the theory includes quantum mechanics. It is also sufficient because we consider only one type of uncertainty, namely, that inherent in quantum mechanics, and not two types of uncertainty, namely, one for quantum effects and another for incomplete information.

The unification of mechanics and thermodynamics is achieved by adding to three fundamental postulates of quantum mechanics (namely, the correspondence postulate, the mean-value postulate, and the dynamical postulate) two more called the energy and stable-equilibrium postulates, which express the implications of the first and second laws of thermodynamics, respectively.

Because the stable-equilibrium postulate appears to be complementary to, consistent with, and independent of, the postulates of quantum mechanics, the second law emerges as a fundamental law of physics that cannot be derived from the other laws.

In contrast to statistical mechanics, the theory is not concerned with "states" that describe outcomes of measurements performed on an ergodic system over long periods of time, or with "states" that describe the subjective knowledge of an observer possessing only partial information about the "actual state" of a system, or with any other type of "state" that does not correspond to identically prepared replicas of a system as defined later. These distinctions among the various definitions of the term state are motivated by important physical considerations that will be touched upon later.

Several theorems that can be derived from the three postulates of quantum mechanics named above have been presented in the literature. One of these is that to every state of a system specified by means of a given preparation there corresponds a Hermitian operator $\hat{\rho}$, called the density operator, which is an index of measurement statistics. The incorporation of the stable-equilibrium postulate into the theory, however, gives rise to additional theorems that are new to quantum physics. Some of these new theorems are as follows:

1.   The maximum energy that can be extracted adiabatically from any system in any state is solely a function of the state. In general, it is smaller than the energy with respect to the ground state, and it can be even zero.

2.   For any state of a system, nonequilibrium, equilibrium, or stable equilibrium, a property S exists that is proportional to the total energy of the state minus the maximum energy that can be extracted adiabatically from the system in combination with a reservoir.

3.   For statistically independent systems, the property S is extensive, it is invariant during all reversible adiabatic processes, and it increases during all irreversible adiabatic processes.

4.   Property S is proportional to $\mathrm{Tr}(\hat{\rho} \ln \hat{\rho})$, where "Tr" denotes the trace of the operator that follows.

5.   The necessary and sufficient condition for stable equilibrium is that S should be at its maximum value for fixed expectation values of energy, numbers of particles of species, and external parameters.

6.   The only equilibrium states that are stable are those for which the density operator yields the canonical distribution if the system is a petit system, and the grand canonical distribution if the system is a grand system.

7.   Property S is defined as the entropy of any state because: (a) for stable equilibrium states, S is shown to be identical to the entropy of classical thermodynamics; and (b) for any state, theorems 2, 3, and 5 above are also theorems of classical thermodynamics.

8.   Classical thermodynamics is an exact but special theory resulting from the application of the present unified theory to systems passing through stable equilibrium states. The present theory in general and, therefore, classical thermodynamics in particular apply regardless of whether the system has a small or a large number of degrees of freedom, and regardless of whether the system is small or large in size.

An idea that is believed to be original with the present theory is that the second law, expressed here in the form of the stable equilibrium postulate, implies that systems may be found in mixed states characterized by irreducible uncertainties, i.e., uncertainties that cannot be represented by a mixture of pure states. These uncertainties are associated with the particles or, more generally, the degrees of freedom of the system. Conversely, the second law is a manifestation of irreducible uncertainties associated with mixed states of a system, and, therefore, with the constituent particles of the system.

The interrelation between irreducible quantal uncertainties and the maximum energy that can be extracted adiabatically from a system represents a radical departure of the present work from other statistical theories, classical or quantum.

In what follows, we present the postulates and some theorems without proofs. Throughout the presentation we assume that the reader is familiar with quantum theory and thermodynamics, and emphasize only points of special relevance to the unified theory.

Postulate 1: Correspondence Postulate. Some linear Hermitian operators on Hilbert space which have complete orthonormal sets of eigenvectors (eigenfunctions) correspond to physical observables of a system. If operator $\hat{P}$ corresponds to observable $\overline{P}$, then operator $F(\hat{P})$, where F is a function, corresponds to observable $F(\overline{P})$.

Postulate 2: Mean-Value Postulate. To every ensemble of measurements performed on identically prepared replicas of a system there corresponds a real linear functional $m(\hat{P})$ of the Hermitian operators $\hat{P}$ of the system such that if $\hat{P}$ corresponds to an observable property $\overline{P}$, $m(\underline{P})$ is the arithmetic mean P of the results of the ensemble of $\overline{P}$-measurements.

Theorem. For each of the mean-value functionals $m(\hat{P})$ there exists a Hermitian operator $\hat{\rho}$ such that for each $\hat{P}$ the following relations hold:

$$P = m(\hat{P}) = Tr(\hat{\rho}\hat{P}) = Tr[\rho][P] \tag{1}$$

where "Tr" denotes the trace of the operator or matrix that follows, and "[x]" the matrix of operator $\hat{x}$. The operator $\hat{\rho}$ is known as the density operator, and $[\rho]$ as the density matrix.

Postulate 3: Energy Postulate. The energy E of a system is defined by the relation

$$E = Tr(\hat{\rho}\hat{H}) = Tr[\rho][H]$$

where $\hat{H}$ is the Hamiltonian operator of the system. It is conserved for all processes in an isolated system.

Theorem. For any physically realizable preparation of a separable system subject to fixed parameters and corresponding to density operator $\hat{\rho}$, the probability $W(P_m)$ that a $\overline{P}$-measurement will yield the eigenvalue P_m of the operator $\hat{P}$ is given by the relation

$$W(P_m) = Tr(\hat{\rho}\hat{P}_m) = Tr[\rho][P_m] \qquad (2)$$

where $\hat{P}_m$ is the projection operator onto the subspace $\hat{P}$ belonging to eigenvalue P_m.

Theorem. The only possible results of $\overline{P}$-measurements are the eigenvalues P_m of $\hat{P}$ for all m, where $\hat{P}$ is the operator corresponding to property $\overline{P}$.

Representation of State. By virtue of its features, the density operator $\hat{\rho}$ is the index of the measurement statistics of quantum physics. It will be seen from Postulate 4 below that $\hat{\rho}$ is also the seat of causality for certain types of changes of state. In addition, it will be shown that, for an ensemble of identically prepared replicas of a system, $\hat{\rho}$ is irreducible; i.e., the ensemble cannot be subdivided into subensembles each of which would yield upon measurement statistics different from the statistics of $\hat{\rho}$. Alternatively, for an ensemble of identically prepared replicas of a system, $\hat{\rho}$ corresponds to irreducible uncertainties of measurement results.

The statements and predictions of our theory apply only to ensembles of identically prepared systems. For such ensembles, the density operator $\hat{\rho}$ may be used to represent the state of the system. This representation gives explicit recognition to the idea that the theory must be confined for the most part to assertions as to the probability that a measurement on a system will yield a particular eigenvalue. Thus, a state will not usually specify the result of each measurement that can be performed on the system in that state.

Postulate 4: Dynamical Postulate.
(1) Any two states of a system that are interconnected by a physical process can always be interconnected by means of one or more reversible processes.
(2) For every system, reversible separable processes always exist for which the temporal development of the density operator $\hat{\rho}$ is given by the relation

$$\hat{\rho}(t_2) = \hat{T}(t_2, t_1) \, \hat{\rho}(t_1) \, \hat{T}^+(t_2, t_1) \qquad (3)$$

where $\hat{T}(t_2, t_1)$ is a unitary operator in time (the evolution operator), and $\hat{T}^+(t_2, t_1)$ is the Hermitian conjugate of $\hat{T}(t_2, t_1)$.

When the Hamiltonian operator $\hat{H}$ of the system is time independent because the values of the parameters are fixed, then the unitary operator $\hat{T}(t, t_1)$ is given by the relation

$$\hat{T}(t, t_1) = \exp\left[-(2\pi i/h) \, \hat{H}(t - t_1)\right] \qquad (4)$$

whereas, when the Hamiltonian operator $\hat{H}(t)$ is an explicit function of time because the values of the parameters are variable, then $\hat{T}(t, t_1)$ conforms to the relation

$$\partial \hat{T}(t, t_1)/\partial t = -(2\pi i/h) \hat{H}(t) \hat{T}(t, t_1) \qquad (5)$$

where h is Planck's constant.

Postulate 5: Stable-Equilibrium Postulate. Any independent
separable system subject to fixed parameters has for each set of
(expectation) values of energy and numbers of particles of constituent species a unique stable equilibrium state.

This postulate brings into our theory the essence of the
second law of thermodynamics. In fact it has been used by
Hatsopoulos and Keenan (8) as a form of the second law in the development of classical thermodynamics. Its introduction along
with the four other postulates results in a theory that embraces
the principles of thermodynamics in addition to those of quantum
mechanics with a single physical meaning of the term state.

The stable-equilibrium postulate does not preclude the existence of many equilibrium states for given values of parameters and
for given expectation values of energy and numbers of particles.
Because any state that satisfies the relation $\hat{\rho}\hat{H} = \hat{H}\hat{\rho}$ could be an
equilibrium state, such states are numerous. The postulate
asserts, however, that, among the many equilibrium states that can
exist for each set of values of parameters, energy, and numbers
of particles, one and only one is stable.

This postulate applies to all systems regardless of size or
numbers of degrees of freedom, including systems having only one
degree of freedom. Of course, the validity of classical thermodynamics for stable equilibrium states of systems with a small number of degrees of freedom was emphasized by Gibbs and others.

It will be shown that the stable-equilibrium postulate re-
stricts application of the theory to states defined by irreducible
uncertainties.

Theorem. Starting from a stable equilibrium state, a separable system cannot do work in any adiabatic process involving
cyclic changes of parameters (CCP process).

In effect this theorem denies the existence of a perpetual
motion machine of the second kind (PMM2), namely, a device acting
as a "Maxwellian demon."

Theorem. From any state of a system, the maximum energy that
can be extracted adiabatically in a CCP process is the work done
in a reversible adiabatic process that ends in a stable equilibrium state. Moreover, the energy change of a system starting
from a given state and ending at a stable equilibrium state is the
same for all reversible adiabatic CCP processes. We call this
energy the adiabatic availability.

Theorem. For any system in any state, a property S exists
that remains invariant in any reversible adiabatic process, that
increases in any irreversible adiabatic process, and that is

additive for independent separable systems. Moreover, the only expression that satisfies the requirements for S and fits experimental data is

$$S = -k \text{ Tr}(\hat{\rho} \ln \hat{\rho}) = -k \text{ Tr}[\rho] \ln [\rho] \qquad (6)$$

where k is the Boltzmann constant.

   Theorem. For any process experienced by an independent separable system having fixed values of energy, numbers of particles, and parameters (namely, a process of an isolated system), the quantity S must either increase or remain invariant.

$$(DS)_{isol} \geqslant 0 \qquad (7)$$

Relation (7) represents the principle of nondecrease of S.

Graphical Representations

   Because a state can be defined by the values of its independent properties, states can be represented by points in a multidimensional property space. In general, the graphical representation is unwieldy because the number of independent properties of a given state can be very large. Nevertheless, useful information often can be summarized by a projection of the multidimensional property space on a two-dimensional plane. One such plane is the $\text{Tr}(\hat{\rho}\hat{H})$ vs. $[-k \text{ Tr}(\hat{\rho} \ln \hat{\rho})]$ plane, namely the E vs. S plane.

   Given a system having fixed numbers of particles (dispersion-free or not) and fixed parameters, the projection of property space on the E-S plane has the shape of the cross-hatched area shown in Figure 1. Each point in this area represents a large number of states having the same values E and S, except for points along the curve $E_g A_0 A_0'$, each of which represents one and only one state.

   For the given values of numbers of particles and parameters, and for values of energy greater than the ground-state energy E_g, the boundary $E_g E_1$ at S = 0 corresponds to all the pure states of the system, namely, to all states that can be described quantum mechanically by wave functions or idempotent matrices. Thus, pure-state quantum mechanics is zero-entropy physics.

   For the given fixed values of numbers of particles and parameters, the curved boundary $E_g A_0 A_0'$ in Figure 1 represents the stable-equilibrium-state relation E vs. S. Its shape is concave as shown because $(\partial E / \partial S)_{n,\beta}$ is an escaping tendency for energy. It reflects the following results of our theory: (a) For each value S_1 for the entropy, stable equilibrium state A_0 is the state of minimum energy; (b) for each value E_1 of energy, stable equilibrium state A_0' is the state of maximum entropy; (c) because each stable equilibrium state is unique, the temperature $(\partial E / \partial S)_{n,\beta}$ is uniquely defined at each point of $E_g A_0 A_0'$; and (d) the ground state

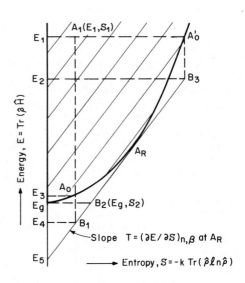

*Figure 1.   Projection of property space
on the energy vs. entropy plane*

is nondegenerate and corresponds to S = 0 and T = 0. The nonde-
generacy of the ground state is a consequence of the third law of
classical thermodynamics. The boundary $E_gA_0A_0'$ represents the
stable equilibrium states of the system, which may be treated by
classical thermodynamics. Thus, stable-equilibrium-state quantum
mechanics is constrained-maximum-entropy physics.

Starting from a state on the boundary $E_gA_0A_0'$ of given energy
E_1, a Maxwellian demon would allow the system to do work only and,
therefore, bring it to a state of energy $E_2 < E_1$. But Figure 1
shows that such a process necessarily implies a decrease of
entropy, which is impossible.

For a given state A_1 (Figure 1), the energy $E_1 - E_3$ is the
adiabatic availability of A_1. In general, it is seen from the
figure that the adiabatic availability varies from $E_1 - E_g$ for a
pure state of energy E_1, to zero for the stable equilibrium state
A_0' corresponding to E_1, depending on the entropy of the state.
This limitation on the amount of work that can be extracted from a
system with no net change in parameters results from the stable-
equilibrium postulate. Although it cannot be derived from the
laws of quantum mechanics, it compares favorably with them in
scientific validity.

For a given reservoir R at temperature T, a line of slope T
can be drawn tangent to $E_gA_0A_0'$ as shown in Figure 1. The point
of tangency A_R is the stable equilibrium state of the system in
question that has a temperature $(\partial E/\partial S)_{n,\beta}$ equal to T. For a
given state A_1 it can be readily verified that the energy $E_1 - E_4$
is the adiabatic availability of A_1 of the system in combination
with reservoir R. It is seen from the figure that this availability varies from a maximum $E_1 - E_5$ for a pure state of energy
E_1, to a minimum $E_1 - E_2$ for the stable equilibrium state A_0' corresponding to E_1, depending on the entropy of the state. These
availabilities of states with values E_1 and $S < S_2$ are greater
than the energy $E_1 - E_g$ of the system above the ground state.

Heat interactions are represented in Figure 1 by paths that
follow the stable-equilibrium-state curve $E_gA_0A_0'$. For these interactions, and for these only, the amount dE of energy transferred is uniquely related to the amount dS of entropy transferred,
namely, $dE = \delta Q = T\,dS$. For end states within the cross-hatched
area, neither is T definable nor can a unique dS be associated
with a given amount of energy transfer dE. It follows that non-
adiabatic interactions, in general, are not heat interactions.

In view of these results, entropy can be used as a measure of
dispersions. Pure states (idempotent density matrices) have zero
entropy, whereas stable equilibrium states have an entropy larger
than that of any other state with the same values of energy,
parameters, and number of particles.

On the Meaning of State

Here we present the precise definition of a state (pure or
mixed) that is subject to the predictions of the unified theory.

The dominant theme of quantum theory is that its causal statements about a system are probabilistic. In other words, the epistemic rule of correspondence, which relates experience to quantum-theoretical states, involves probabilistic concepts in an essential way. In particular, an essential premise of quantum theory is that the physical condition or state of a system at a given time cannot be fully disclosed experimentally unless many measurements are made on replicas of the system prepared in a specified manner. Conversely, an inherent prerequisite of quantum theory is that a preparation of a system be specified and uniquely associated with a state prior to any attempt to reveal experimentally the characteristics of the state. It is this prerequisite that clearly distinguishes quantum mechanics from classical mechanics. It has been discussed extensively in the literature.

The index of measurement statistics corresponding to a given preparation can be expressed in the form of a density operator $\hat{\rho}$. Some preparations result in states described by density operators that are pure (density matrices are idempotent), and some in states described by density operators that are mixed (density matrices are not idempotent). In the context of the quantum mechanical postulates, the preceding sentence is all that need be said about any given preparation and, therefore, any given state.

It is frequently stated that a mixed density operator refers to an ensemble made up of systems each of which is in a pure state. Such a statement, as pointed out by Park (9), is meaningless. In quantum theory, the only experimentally observed reality is that which is revealed by the statistics of measurements performed on an ensemble of identical systems prepared in a specified manner. If a given preparation results in a mixed density operator, then this operator represents the only meaningful reality of the state. Park points out that a general quantum ensemble characterized by a density operator $\hat{\rho}$ can be numerically (as opposed to operationally) subdivided in an infinite variety of ways into pure or mixed subensembles, namely,

$$\hat{\rho} = \sum_{k} w_{k}\hat{\rho}_{k} \quad \text{and} \quad \sum_{k} w_{k} = 1 \qquad (8)$$

where $\hat{\rho}_{k}$ is pure or mixed and $0 < w_{k} < 1$ for all k.

On the other hand, we may raise a different question: Is it possible to extablish an operationally meaningful criterion that will distinguish between (a) preparations resulting in dispersions that are due partially to nonquantum effects (or to lack of knowledge) and partially to quantum effects and (b) preparations resulting in dispersions that are solely due to quantum effects? The answer to this question is yes. Prior to presenting the criterion, however, we give an explicit operational definition of the term identically prepared systems.

Definition of Unambiguous Preparation

We shall define a preparation as unambiguous and the resulting ensemble as consisting of identically prepared systems that
are in a state $\hat{\rho}$ and that are subject to the predictions of the
present unified theory if and only if the subdivision of the ensemble prior to measurement into two or more subensembles, according to any conceivable operational rule, will always result in
each subensemble being in the same state $\hat{\rho}$; in other words, the
statistics of measurements performed on any subensemble after subdivision will be representable by the same density operator $\hat{\rho}$ as
the statistics of the overall ensemble.

For example, consider an ensemble with its members numbered
consecutively. Suppose that measurements are made on the subensemble consisting on, say, all even-numbered members, and on the
subensemble consisting of all the odd-numbered members. If the
probabilities that are derived from the measurements in the first
subensemble are identical to the probabilities that are derived
from the second subensemble, and this identity obtains for any
conceivable subdivision of the original ensemble into subensembles,
then the preparation is unambiguous. If measurements performed
on the subensembles after subdivision yield statistics that are
represented by density operators that are different than that of
the overall ensemble, the preparation will be called ambiguous.

These definitions are motivated by the stable-equilibrium
postulate, and their importance will become evident from the subsequent discussion.

An Analogy from Probability Theory

The concept of an unambiguous preparation may be illustrated
by means of a simple example from probability theory. Suppose
that we cut a large number of metallic rods each appearing to have
the same length, and that we wish to verify through measurements
if indeed the cutting process results in identical lengths. Suppose further, however, that because of either the available
measurement technique or some inherent characteristics of the rods,
or both, the results of the measurements include a random but
statistically unique error so that, even if all the rods were cut
to identical lengths, the measurement results would be dispersed.
Under these conditions, the question arises: Is is possible to
determine whether the rods were prepared by the same cutting procedure?

We may answer this question by proceeding as follows. First,
we measure the lengths of the set of all the rods, make a graph
of frequency vs. length, and find the average length. Next we
divide the rods into two subsets: one consisting of the rods
having measured lengths longer than the average, and the other
consisting of the rods having lengths shorter than the average.
Then we repeat the length measurements and make frequency vs.

length graphs for each subset. Elementary probability theory indicates that, if indeed all the rods were prepared by the same cutting procedure and the observed dispersions were solely due to random effects not associated with the cutting procedure, then the frequency graphs corresponding to the two subsets would be identical. On the other hand, if the rods were not prepared by the same cutting procedure, then the frequency graphs of the two subsets would not be identical.

Theorem - Criterion. Given an ensemble of systems prepared by a preparation Z and consisting of several subensembles, the preparation is unambiguous if: (a) measurements performed from time to time on each system of the ensemble and on each system of the subensembles yield results that are statistically independent; and (b) the joint probabilities for such results are the same for both the ensemble and the collection of the subensembles.

Theorem - Criterion. Given an ensemble of identical systems having a Hamiltonian operator $\hat{H}$ and a density operator $\hat{\rho}$, and consisting of two or more subensembles each of which is prepared by means of an unambiguous preparation, the entropy defined in terms of availability is either equal to $-k\,\mathrm{Tr}(\hat{\rho}\,\ln\,\hat{\rho})$ if the preparations of the subensembles are identical, or smaller than $-k\,\mathrm{Tr}(\hat{\rho}\,\ln\,\hat{\rho})$ if the preparations of the subensembles are different.

On Irreducible Dispersions

The criteria for unambiguous preparations given above provide operational means for distinguishing between dispersions of measurement results that are inherent in the nature of a system and those that are related to voluntary or involuntary incompleteness of experimentation. The former represent characteristics of a system that are beyond the control of an observer. They cannot be reduced by any means, including quantum mechanical measurement, short of processes that result in entropy transfer from the system to the environment. For pure states, these irreducible dispersions are, of course, the essence of Heisenberg's uncertainty principle. For mixed states, they limit the amount of energy that can be extracted adiabatically from the system.

Additional dispersions introduced by voluntary or involuntary incompleteness of experimentation represent inadequacies in the knowledge of observers. As such, though subject to improvement, they are not subject to the full prescriptions of the laws of physics.

The existence of irreducible dispersions associated with mixed states is required by Postulate 5, which expresses the basic implications of the second law of classical thermodynamics. Alternatively, the present work demonstrates that the second law is a manifestation of phenomena characteristic of irreducible quantal dispersions associated with the elementary constitutents of matter.

The possibility of a relation between the second law (in the form of the impossibility of a Maxwellian demon) and irreducible dispersions associated with pure states (represented by Heisenberg's uncertainty principle) was suggested by Slater (10). His suggestion was not adopted, however, because Demers (11) proved that dispersions associated with pure states are insufficient to account for the implications of the second law, especially with regard to heavy atoms at low pressures. In the present work, we can relate the second law to quantal dispersions of mixed states because we have accepted the existence of dispersions of mixed states that are irreducible.

In conclusion, in the unified theory the state of any system is described by means of probabilities that are inherent in the nature of the system and that are associated with measurement results obtained from an ensemble of systems of unambiguous preparation. Moreover, the second law of thermodynamics emerges as a fundamental law related to irreducible quantal dispersions of mixed states and applicable to systems of any size, including a single particle.

A key element of the theory is the statement of operational criteria for the distinction between ambiguous and unambiguous preparations (pure or mixed).

For unambiguous preparations, the theory reveals limitations on the amount of work that can be done by a system adiabatically and without net changes in parameters. These limitations are due to irreducible dispersions inherent in the state of the system. They are maximal when the dispersions correspond to a stable equilibrium state.

Literature Cited

1. Shannon, C.E., Bell System Tech. Jl., Vol. 27 (1948).
2. Shannon, C.E. and Weaver, W., "Mathematical Theory of Communication," Univ. of Illinois Press: Urbana, 1949.
3. Jaynes, E.T., Phys. Rev., Issue 4, Vol. 106 (1957); and Phys. Rev., Issue 2, Vol. 108 (1957).
4. Katz, A., "Principles of Statistical Mechanics," W.H. Freeman: San Francisco, 1967.
5. Szilard, L., Z. Physik, Vol. 32, p. 753 (1925).
6. Brillouin, L., "Science and Information Theory," Academic Press: New York, 1956.
7. Hatsopoulos, G.N. and Gyftopoulos, E.P., Foundations of Physics, Part I, Issue 1, Vol. 6, p. 15 (1976); Part IIa, Issue 2, Vol. 6., p. 127 (1976); Part IIb, Issue 4, Vol. 6, p. 439 (1976); Part III, Issue 5, Vol. 6, p. 561 (1976).
8. Hatsopoulos, G.N. and Keenan, J.H., "Principles of General Thermodynamics," John Wiley: New York, 1965.
9. Park, J.L., Am. Jl. Phys., Vol. 36, p. 211 (1968).
10. Slater, J.C., "Introduction to Chemical Physics," McGraw-Hill: New York, 1939.
11. Demers, P., Can. J. Res., Vol. 22, p. 27 (1944); Vol. 23, p. 47 (1945).

RECEIVED November 1, 1979.

The Information Theory Basis for Thermostatics:
History and Recent Developments

MYRON TRIBUS

Center for Advanced Engineering Study, Massachusetts Institute of Technology, Cambridge, MA 02139

When Rudolf Clausius (1) formalized the laws of thermodynamics in 1850, he started a thermodynamic tradition. Clausius based his work on the writings of Rumford, Joule, Carnot, and Mayer. The approach is called "classical." Its features are indicated in Figure 1 which suggests that the mental constructs of classical thermodynamics are quite close to observables. The concepts are macroscopic, the relations are operational, and the predictions are deterministic. They are derivable and defined by instruments, experimental conditions, and the results of experiments (Figure 1).

In parallel to this classical approach another view was developed by equally illustrious contributors. Their work was much less unified and known under names such as "kinetic theory" and "statistical mechanics."

The second approach was more abstract. It invoked additional constructs not readily accessible to our senses and not readily related to everyday experience. Electrons, photons, phonons, holes, bonds, energy states, etc., are the basic constructs of the other approach. Figure 2 gives a view of how these approaches are related.

Microscopic approaches have scored many notable successes, including the entire worlds of chemistry, nuclear power, and solid state electronics. To those who are very much concerned with the logical and philosophical foundations of things, the logical untidiness of micro approaches has been a bit of an embarrassment. It is indeed a brilliant accomplishment to deduce the second law in the style of Carnot, but the accomplishments in electronics in developing, say, a theory of amorphous semiconductors are also impressive even if the theory seems less firmly grounded.

Some academics have been troubled by the fact that blending micro- and macro- views has not gone smoothly, especially in thermodynamics. After all, they often stand before students and explain where the basic ideas come from. It is discomforting, to say the least, to have the most brilliant students turn off because they sense difficulties being glossed over.

0–8412–0541–8/80/47–122–275$05.00/0
© 1980 American Chemical Society

```
MACRO
                          Definitions    Relations
"IMAGERY"  (Construct)    Concepts    -  Operations  - Predictions


"REALITY" (Observation)  Instruments  -  Experiments - Results
```

Figure 1.   Features of the "classical" approach to the laws of thermodynamics

```
                        Microscopic  _  Relations   _  Microscopic
                        Definitions     Operations     Predictions
              MICRO
           Constructs    Concepts

"IMAGERY" - - - - - -↑(Divide) - - - - - - - - - -↓(Average)

                        Macroscopic  _  Relations   _  Macroscopic
                        Definitions     Operations     Predictions
              MACRO
           Constructs    Concepts


"REALITY"(Observation) Instruments   -  Experiments - Results
```

Figure 2.   View of how various approaches to the laws of thermodynamics are related

to say the same thing--or were they merely superficial analogs?

Only three years after Shannon's paper, Jerome Rothstein published the answer to the dilemma (10). His paper is only four paragraphs long. It omits mathematics, but answers completely every question about the connection between the two fields. Being philosophical and qualitative and, at the time speculative, the paper was not widely read and appreciated.

Nine years after Shannon's paper, Edwin T. Jaynes published a synthesis of the work of Cox and Shannon (11). In this paper Jaynes presented the "Maximum Entropy Principle" as a principle in general statistical inference, applicable in a wide variety of fields. The principle is simple. If you know something but don't know everything, encode what you know using probabilities as defined by Cox. Assign the probabilities to maximize the entropy, defined by Shannon, consistent with what you know. This is the principle of "minimum prejudice." Jaynes applied the principle in communication theory and statistical physics. It was easy to extend the theory to include classical thermodynamics and supply the equations complementary to the Rothstein paper(12).

Shannon, Cox, and Jaynes introduced a new layer in our diagram of the logic. As indicated in Figure 3, the question was no longer how to justify micro- and macro- concepts and their relation to one another. Rather, the question became: Given the incomplete information, how can we design useful concepts and relations? How do we design mental constructs to meet our purposes?

Figure 3 is admittedly an inexact representation, but it does show the main point that the resolution of the relation between the macro and micro views required a new set of ideas about the imaginal processes and not about the micro and macro concepts themselves.

As far as I can tell by talking with contemporary thermo-dynamicists, especially those who grew up with the traditions of classical thermodynamics, these revolutionary ideas have had very little effect on them. But the impacts of the Shannon and Jaynes' papers on others has been most dramatic. A few months ago I ordered a computer search of one particular data base. We looked for all papers published between 1970 and 1975 in which Shannon or Jaynes or both appeared as references. There were over 400 literature citations in such fields as systems theory, biology, neurology, meteorology, statistical mechanics, thermo-dynamics, irreversible processes, reliability, geology, psychiatry, communications theory and even urban studies, transportation and architecture.

In May of 1978, we held a conference at M.I.T. dealing with the Maximum Entropy Principle. Papers were given in a variety of fields illustrating how widely these influences have spread. The papers have been published by The M.I.T. Press under the title, "The Maximum Entropy Formalism," 1978.

	Principles	Design	Principles	
	in	of		
"IMAGINAL	Logic	Functions	(Maximum	
PROCESSES"	Logical	for	Entropy)	Methods
	Constructs	Specified Purposes		
	(Probability)	(Information)	(Entropy)	

	MICRO			
	Constructs	Definitions	Relations	
"IMAGERY"	Concepts	and	and	Predictions
	Abstract	Concepts	Operations	

- - - - - - - - - - - - - ↑ (Divide) - - - - - - - - - - ↓ (Average)

	MACRO			
	Constructs	Definitions	Relations	
"IMAGERY"	Concepts	and	and	= Predictions
	Abstract	Concepts	Operations	

	Observations			
"REALITY"	Perceptions	Instruments	Experiments	Results
	Concrete			

Figure 3

In the remainder of these remarks I shall confine my attention to applications in thermodynamics and irreversible processes.

In thermodynamics (or more accurately, thermostatics) the principal contribution of information theory is to redefine the basic ideas of classical thermostatics along the lines forecast by Rothstein. The first quantitative treatment was published in 1961, followed by a textbook and later a sequence of papers (12, 13, 14, 15, 16, 17, 18, 19, 20, 21).

The essential logical steps are presented here in as compact and logical form as I know how.

We presume an observer who has available various measuring instruments which average out (in some way) the influences of the tiny particles which make up a system. The only information the observer has about the system is obtained from instruments.

The information the observer has about systems in general, we denote by "X." "X" includes knowledge that matter is composed of particles of various kinds and lifetimes, that individual particles are not distinguishable, that in their states of motion they are discrete (i.e., do not exhibit continuous variation) and that Newtonian and Quantum Mechanics both ascribe a constant of the motion called "energy." "X" tells our observer that all systems, in principle, have quantum states which are denumerable, even if they are infinite in number, and that from instruments alone it is not possible to say in which quantum state the system actually resides.

Thus, our observer is in the typical position of humans who do not know everything but they do know something and, therefore, if they wish to be honest, need a way of communicating neither more nor less than they really know.

A theory is a connected set of constructs. Our observer sets out to design a suitable theory. The observer knows about the principle of conservation of energy (not to be confused with the first law) and realizes people will wish to compute the energy of a system. Since the information from instruments is partial, they will not be able to compute energies with precision since the available information is incomplete.

Our observer knows from the work of Cox that the only rational way to communicate incomplete information is via the probability function. So the observer writes

$$p_i \equiv p\left(A_i \mid X\right) \tag{3}$$

to mean p_i is the probability that the system is in state i conditional on knowledge X. Following Cox and Jaynes, our observer doesn't equate p_i to a "state of things"--it is a number assigned to represent a state of knowledge, "X," to be precise. The only constraints our observer sets on the set $\{p_i\}$ are

$$\sum p_i = 1 \tag{4}$$

$$\sum_i P_i \, \varepsilon_i = <\varepsilon> \tag{5}$$

Note--our observer doesn't know the expectation energy ($<\varepsilon>$). What the observer does know is that the system is in some state (Equation 4) and if it is stable, it exhibits a reproducible energy (Equation 5).

To assign values to the probabilities, our observer makes use of the principle of minimum prejudice as described by Jaynes: "Assign to the p_i values, consistent with the

constraints, which maximize the entropy, S_I." By well-known methods, the observer finds

$$P_i = e^{-\psi - \beta \varepsilon_i} \tag{6}$$

From this result, without any further attempts at physical interpretation, mathematical manipulations yield:

$$S_I = k\psi + k\beta <\varepsilon> \tag{7}$$

$$\psi = \ln \sum_i e^{-\beta \varepsilon_i} \tag{8}$$

$$<\varepsilon> = -\partial \psi / \partial \beta \tag{9}$$

Among other things our observer's "X" tells that the energy associated with a quantum state, i, depends on the volume of the system, V and (if all other external fields are constant)

$$\varepsilon_i = \varepsilon_i \, (V) \tag{10}$$

from which   (Note: p = probability, P = pressure)

$$P_i \equiv -\partial \varepsilon_i / \partial V \tag{11}$$

from (10) and (8),

$$\psi = \psi \, (\beta, V) \tag{12}$$

$$d\psi = \frac{\partial \psi}{\partial \beta} \, d\beta + \frac{\partial \psi}{\partial V} \, dV \tag{13}$$

$$= -<\varepsilon> \, d\beta + \frac{\partial \psi}{\partial V} \, dV \tag{14}$$

from (7)

$$dS_I = kd\psi + k\beta d<\varepsilon> + k<\varepsilon>d\beta \tag{15}$$

and (14)

$$dS_I = k\beta d<\varepsilon> + k \frac{\partial \psi}{\partial V} dV \tag{16}$$

since

$$d<\varepsilon> = \sum \varepsilon_i \, dp_i + \sum p_i \, d\varepsilon_i \tag{17}$$

$$= \sum \varepsilon_i \, dp_i + \sum p_i \frac{d\varepsilon_i}{dV} dV \tag{18}$$

use Equation 11,

$$d<\varepsilon> = \sum \varepsilon_i \, dp_i - \sum p_i \, P_i \, dV \tag{19}$$

$$= \sum \varepsilon_i \, dp_i - <P> \, dV \tag{20}$$

since from Equation 1,

$$dS_I = -k \sum (\ln p_i + 1) \, dp_i \tag{21}$$

and from (6)

$$dS_I = -k \sum (1 - \psi - \beta\varepsilon_i) \, dp_i \tag{22}$$

$$= k\beta \sum \varepsilon_i \, dp_i \tag{23}$$

from (20) and (23)

$$dS_I = k\beta \, d<\varepsilon> + k\beta \, <P> \, dV \tag{24}$$

If we recognize that when dV = 0, the increment d<ε> is what is called heat, the identity of S_I with S_T is established.

There are many subtle details to be considered--they have been treated in many publications already. The important point to keep in mind is that in eliminating Equations 6 and 8, our observer has developed an entirely macroscopic theory. Only Equations 1, 6 and 8 involve micro concepts. Thus, using the Jaynes-Cox formalism, our observer has moved downward through the top three levels of Figure 3.

In the space and time available for this presentation, it does not seem useful to demonstrate how terms

$$\beta, \ \psi, \ <P>, \ <P>dV, \ \text{and} \ \sum \varepsilon_i \, dp_i$$

respectively translate into and define the concepts of temperature, Helmholtz free energy, equilibrium pressure, reversible work, and reversible heat. These and many other interesting relations are developed in References 12 to 21. You can also find the same ideas developed in more recent textbooks (22, 23). So we shall say no more here about thermostatics, which is already a mature field, and close with some comments about irreversible processes.

There is only one way for a system to be in equilibrium with its surroundings. There are many ways to be out-of-equilibrium. Therefore, the field of thermostatics is well-defined and mature, and the field of thermodynamics (or irreversible thermodynamics or whatever latest fashion calls it) is less well-defined and growing. By irreversible thermodynamics, we usually mean the study of processes in which spontaneous change is occurring and, therefore, thermostatic entropy is being created. Since, by definition the system is not in equilibrium, the central question is how to relate the entropy change to physical processes other than the definition in Equation 2, since that definition requires "reversible heat transfer."

An interesting example of a non-equilibrium process studied by maximum information entropy methods was reported by Levine. A beam of O is allowed to interact with a beam of CS to produce a beam of CO and S. The O beam is defined by macroscopic constraints and, therefore, the statistics are well-known. The collision between the O and CS produces CO and S. The equations of conservation of momentum and energy do not suffice to describe precisely the final state. There are, in fact, various final states consistent with the known data. Levine uses maximum entropy estimates to describe the initial beams and the final products with excellent agreement between observation and calculation. Levine applies the method to other systems with equally good results.

Jim Keck used maximum entropy methods to treat the very large number of simultaneous equations which occur when we try to calculate the pollutants produced in minute quantities during combustion (25). Faced with a large number of rate equations (say 20 or more) for which rate constants were not available, Professor Keck relied upon the maximum entropy principle to make the best possible estimate, again with good results.

But why should maximum entropy estimates work? Why is it that if we declare there are certain bounds on our knowledge, we end up with excellent predictive power? The answer is straight-forward: When we do an experiment, we exert control over some variables and not others. Within the limits imposed, the system does whatever it does. The maximum entropy estimate is developed to reflect these same constraints. If the maximum entropy estimate agrees with the observed behaviour, we conclude it is a good way to describe the system. If there is disagreement, we look for the presence of additional constraints in the experimental setup which were not taken into account in the maximum entropy estimate. Since we are dealing only with repeatable experiments, i.e., experiments which yield essentially the same result, even though all variables are not tightly controlled, we are assured that our attempts to make predictions based on

incomplete data are not foolish. When we encounter a class of
problems, such as is thermostatics, where only a few parameters
need be controlled (volume, expected energy, expected composition)
to achieve a reproducible end state (equilibrium), we are not
surprised if this end state corresponds to maximum entropy on
either S_I or S_T and at that point $S_I \equiv S_T$.

Literature Cited

1. Clausius, Rudolf. "The Mechanical Theory of Heat" translated
by Hirst, Van Voorst, London, 1867.

2. Maxwell, James, C. "A Theory of Heat"; 1871; p. 328.

3. Szilard, L. Uber die Entropie verminerung in einem
thermodynamischen System bei Eingriffen intelligenter Wesen,
Z. _Physik_, 1929; _53_, 840.

4. Van der Waals, J. Uber die Erklarung der Naturgesetze auf
Statisch-Mechanishcher Grundlage, _Physik_. Zerichr. XII,
1911; pp. 547-549.

5. Lewis, G. N. The Symmetry of Time in Physics, _Science_,
June 1970; _71_, p. 569.

6. Gibbs, Josiah W. "Elementary Principles in Statistical
Mechanics," Yale University Press (reprinted) 1948.

7. Shannon, Claude E. A Mathematical Theory of Communication,
The Bell System Technical Journal, 1948; Vol. _27_, pp. 379-623.

8. Tribus, Myron. Thermodynamics--A Survey of the Field, from
the book "Recent Advances in the Engineering Sciences,"
McGraw-Hill Book Co., 1958.

9. Cox, Richard T. Probability, Frequency and Reasonable
Expectation, _Amer. Jour. Phys._, 1946; _14_, 1.

10. Rothstein, Jerome. Information and Thermodynamics, _The
Physical Review_, January 1, 1952; Vol. _85_, No. 1, 135.

11. Jaynes, Edwin T. Information Theory and Statistical
Mechanics. _Phys. Rev._, 1957; _106_, p. 620 and _Phys. Rev._, 1957;
_108_, p. 171.

12. Tribus, Myron. Information Theory as the Basis for Thermostatics and Thermodynamics, _Jour. Appl. Mech._, March 1961;
pp. 1-8.

13. Tribus, Myron. "Thermostatics and Thermodynamics, An
Introduction to Energy, Information and States of Matter,"
D. Van Nostrand, 1961.

14. Tribus, Myron; Evans, Robert B. The Probability Foundation
of Thermodynamics, _Applied Mechanics Reviews_, October 1965;
Vol. _16_, No. 10, pp. 765-769.

15. Tribus, Myron. "Information Theory and Thermodynamics," Boelter Anniversary Volume, McGraw-Hill Book Co., 1963; pp. 348-367.

16. Tribus, Myron; Shannon, Paul T.; Evans, Robert B. Why Thermodynamics is a Logical Consequence of Information Theory, A.I.Ch.E. Jour., March 1966; pp. 244-248.

17. Tribus, Myron. Micro and Macro Thermodynamics. American Scientist, June 1966; Vol. 54, No. 2.

18. Tribus, Myron. Generalizing the Meaning of Heat, Int. Jour. of Heat and Mass Transfer, Pergamon Press, 1968; Vol. II, pp. 9-14.

19. Tribus, Myron; Evans, Robert B. A Minimum Statistical Mechanics from Which Classical Thermostatics May Be Derived, in the book, "A Critical Review of Thermodynamics," Mono Book Co., Baltimore, MD, 1970.

20. Tribus, Myron; McIrvine, Edward C. Energy and Information, Scientific American, September 1971; Vol. 225, No. 3.

21. Tribus, Myron; Costa de Beauregard, Olivier. Information Theory and Thermodynamics--A Rebuttal, Helvetica Physica Acta, 1974; Vol. 47.

22. El-Saden, M. "Engineering Thermodynamics," D. Van Nostrand Co., Princeton, NJ, 1965.

23. Beierlein, Ralph. "Atoms and Information Theory," W. H. Freeman & Co., San Francisco, CA, 1971.

24. Levine, R. D. Maximal Entropy Procedures for Molecular and Nuclear Collisions, proceedings of the conference on Maximum Entropy Formalism, M.I.T. Press, 1979.

25. Keck, James C. Rate Controlled Constrained Equilibrium Method of Treating Reactions in Complex Systems: Maximize the Entropy Subject to Constraints, proceedings of the conference on Maximum Entropy Formalism, M.I.T. Press, 1979.

26. Levine, Raphael D.; Tribus, Myron (editors). "The Maximum Entropy Formalism," The M.I.T. Press, Cambridge, MA, 1978.

RECEIVED October 26, 1979.

INDEX

INDEX